OpenAI API编程实践

(Java版)

[美] 布鲁斯·霍普金斯(Bruce Hopkins) 著

郭 涛 译

清華大学出版社

北 京

北京市版权局著作权合同登记号　图字：01-2024-4439
ChatGPT for Java: A Hands-on Developer's Guide to ChatGPT and OpenAI APIs
by Bruce Hopkins

图书在版编目(CIP)数据

OpenAI API 编程实践 : Java 版 / (美) 布鲁斯·霍普金斯 (Bruce Hopkins) 著 ; 郭涛译. -- 北京 : 清华大学出版社, 2025. 2. -- ISBN 978-7-302-67920-2

Ⅰ. TP312.8

中国国家版本馆 CIP 数据核字第 2025VX8128 号

责任编辑： 王　军
封面设计： 高娟妮
版式设计： 恒复文化
责任校对： 马遥遥
责任印制： 刘　菲

出版发行： 清华大学出版社
网　　址：https://www.tup.com.cn，https://www.wqxuetang.com
地　　址：北京清华大学学研大厦 A 座　　邮　　编：100084
社 总 机：010-83470000　　邮　　购：010-62786544
投稿与读者服务：010-62776969，c-service@tup.tsinghua.edu.cn
质 量 反 馈：010-62772015，zhiliang@tup.tsinghua.edu.cn
印 装 者：北京瑞禾彩色印刷有限公司
经　　销：全国新华书店
开　　本：148mm×210mm　　印　　张：6.125　　字　　数：189 千字
版　　次：2025 年 3 月第 1 版　　印　　次：2025 年 3 月第 1 次印刷
定　　价：59.80 元

产品编号：107455-01

译者简介

郭涛，主要从事人工智能、智能计算、概率与统计学、现代软件工程等前沿交叉领域的研究。出版过多部译作，包括《深度强化学习图解》《机器学习图解》和《Copilot 和 ChatGPT 编程体验：挑战 24 个正则表达式难题》。

译 者 序

在人工智能领域，Transformer 模型的崛起堪称一场革命性的变革。作为一种引领潮流的生成式人工智能模型，Transformer 已成为众人瞩目的焦点。它在语言理解、文本生成等领域展现出的卓越性能，吸引了广泛的关注。随着近年来大规模模型的不断涌现，Transformer 模型更是焕发出新的活力。

目前，全球已有上千个大模型竞相亮相，其中包括 OpenAI 的 GPT 系列、Meta AI 的 LLaMA 系列、Google 的 Gemma、百度的文心一言以及阿里云通义千问等杰出代表。这些大模型不仅具有强大功能，还提供了标准的 API 接口，方便应用开发者调用，为各类应用场景提供了无限可能。

本书主要以 OpenAI 发布的 ChatGPT 为对象，为 Java 程序员而撰写。本书详细介绍了 ChatGPT 的标准接口、调用形式以及丰富的应用场景，重点涵盖了基于 ChatGPT 的结对编程实践、企业 API 调用实战、多模态场景调用以及 AI 智能机器人的实现等内容。本书采用场景案例驱动的方式，结合接口使用说明和代码实战，旨在教会读者如何灵活运用 ChatGPT 接口，满足各种业务场景的需求。

通过阅读本书，读者不仅可了解 ChatGPT 大模型，还可掌握类似的逻辑和思维方式，通过迁移学习方法运用其他大模型，如文心一言、通义千问等。本书篇幅紧凑，内容精辟，面向应用开发人员以及希望利用大模型解决业务场景问题的科学家、工程师。同时，对于非专业人员来说，本书也是一本颇具价值的参考读物。相信通过本书的学习，你将能够更加熟练地运用 ChatGPT 的强大功能，为你的 Java 应用增添新的智慧与活力。

在本书的翻译过程中，我得到了众多人的无私帮助。特别要感谢成都文理学院翻译系的何静老师；作为本书的审校者，她以其深厚的专业知识和严谨的态度，为本书的翻译质量提供了坚实的保障。同时，也要向清华大学出版社的编辑团队表达诚挚的谢意，他们不辞辛劳地进行了大量的编辑与校对工作，确保了本书内容的准确性。

尽管我努力追求翻译的准确性和流畅性，但鉴于原著内容的广度和深度，以及译者自身水平的局限，翻译过程中难免存在不足。因此，我诚挚地邀请各位读者在阅读过程中对发现的任何问题进行批评指正，我将不胜感激。

作者简介

Bruce Hopkins 是技术领域的杰出作家和全球知名专家。他不仅是 Oracle Java 的冠军，还荣获了英特尔软件创新者的称号。此外，他还是 Apress 出版社所出版的 *Bluetooth for Java* 一书的作者，该书展现了他在蓝牙与 Java 技术融合领域的深厚造诣。

审校者简介

Van VanArsdale 是一位在软件行业积累了超过 30 年经验的技术领导者。他拥有马萨诸塞大学洛厄尔分校的计算机信息系统学士学位和密苏里州立大学的计算机信息系统硕士学位。在其职业生涯中，他担任过软件工程师、架构师、经理和教师等多个角色。目前，Van 在一家顶尖的金融服务公司领导着一支高效的团队，并兼任密苏里州立大学的教师，继续为培养新一代的软件人才贡献力量。

序　言

我有幸与 Bruce 相识数年，并亲眼见证了他将最前沿的概念和问题转化为易于理解的语言，让拥有各种背景和能力的程序员都能掌握。Bruce 的丰富经验使得他能够将最新的人工智能技术，如 ChatGPT，巧妙地分解为开发人员在日常工作中可实际运用的核心要素。作为 *Bluetooth for Java* 一书的合著者，他在蓝牙技术仅兴起四年之际便投身其中，展现出了前瞻性的眼光。此外，他还为众多大型技术公司撰写了技术指南，助力开发人员迅速掌握最新技术。

作为微软旗下人工智能公司 Private AI 的联合创始人兼首席执行官，我有机会与全球各地的企业开发人员、经理和高层管理者进行交流。我们讨论的议题不仅限于如何负责任地使用数据，还深入探讨了生成式人工智能(如 ChatGPT)所引发的问题、要处理哪些类型的问题以及从哪里开始等。由于 ChatGPT 等技术相对较新，其底层模型架构 Transformer 直到 2017 年才通过 *Attention Is All You Need* 这篇研究论文问世，因此，这些议题对于管理者和开发者来说都充满了挑战。他们纷纷提出“这项技术能为我带来什么价值？”“我应该在哪些场景中运用它？”“如何入门？”以及“这项技术存在哪些局限性？”等问题。此外，每当新技术问世时，人们总会不可避免地提出质疑：“这一切是否只是炒作？”

值得欣慰的是，本书针对上述问题提供了具体而实用的解答，这对于我们深入理解新技术并挖掘其核心价值至关重要。学习编程语言会让我们以全新的角度思考问题，同样，学习如何有效利用人工智能也会让我们跳出编程的框架，以更广阔的视角审视问题。在自然语言中，我们关注词汇、语法、句法和语义；在编程语言中，我们关注逻辑、数学、语法、规模以及对重要原理的理解；而在人工智能领域，我们更关注数据及其与任务之间的关系。尽管我们可以利用世界上最强大的模型生成文本、导航或从事艺术创作，但如果将它们应用于不合适的任务，结果可能令人失望。然而，如果能够深入理解这些模型的设计初衷，并据此正确使用它们，那么结果定会令人欣喜。

本书通过简明实用的示例，不仅可以帮助你迅速开始使用 ChatGPT 构建项目，还能培养你对这项技术的直觉和洞察力。尽管本书的重点是使用 Java 进行实践，但无论你使用哪种编程语言进行开发，都能从中受益。

Patricia Thaine
Private AI 公司联合创始人兼首席执行官

目　　录

第1章

面向 Java 开发人员的 ChatGPT 简介

1.1 本书读者对象

首先，本书专为那些没有接受过人工智能、自然语言处理、机器学习或深度学习培训的 Java 开发人员而编写。或许你曾听说过“语言模型”这一术语，但在日常工作中，它可能并非你经常使用的概念。

其次，尽管你可能对 ChatGPT 有所耳闻或曾尝试使用过，但对其背后的工作原理可能并不十分了解。同时，你也可能不确定如何开始使用 Java 和 ChatGPT 编写程序，以在你的应用程序和服务中实现相关功能。

注意：

尽管 ChatGPT 已家喻户晓，但其背后的公司 OpenAI 却相对鲜为人知，并未获得广泛的认可。因此，虽然本书主要聚焦于如何在 Java 应用程序中通过编程方式利用 ChatGPT，但实际上我们将使用的 API 的正式名称是 OpenAI REST API。在本书的后续内容中，将交替使用“ChatGPT API”和“OpenAI API”这两个术语。

1.2 本章概述

在本章中，我们将深入解释一些你可能不太熟悉的术语，然后直接进入 ChatGPT Playground。这个 Playground 本质上是一个基于 ChatGPT REST API 的开发者 Web 界面，旨在帮助开发者更好地理解和使用 ChatGPT。因此，当你熟悉 Playground 的操作后，调用和响应 ChatGPT API 将变得轻而易举。

1.3 立即下载代码

本书附赠了丰富的示例和示例代码，让你不必再费心手动输入。为便于使用，我们提供了用于访问源代码的 GitHub 代码仓库的 URL，网址为 https://github.com/Apress/ChatGPT-for-Java。

1.4 那么，ChatGPT 究竟是什么，为什么我需要使用 OpenAI API?

简单而言，ChatGPT 是 OpenAI 开发的一组语言模型(是深度学习算法训练的结果)，能够根据给定的提示生成类似人类的文本响应。那么什么是提示？提示就是你向 ChatGPT 提供的输入文本。因此，我们先给 ChatGPT 一个提示，询问它对 Java 设计模式的熟悉程度(代码清单 1-1)。

代码清单 1-1 提示。在 ChatGPT 中给出有关设计模式提示

```
Describe 3 commonly used design patterns used by Java
developers and why those patterns are important
```

中文注释

描述 Java 开发人员常用的 3 种设计模式，以及这些模式的重要性

如果你有机会在一家专业使用 Java 的公司工作，那么你可能在面试中遇到过与上述提示非常相似的问题。代码清单 1-2 是 ChatGPT 的响应。

代码清单 1-2　响应。ChatGPT 对代码清单 1-1 中提示的响应

```
    Certainly! Design patterns are reusable solutions to common problems
that arise in software design. They are best practices that experienced
developers have discovered and refined over time. Java developers
frequently utilize design patterns due to the language's object-oriented
nature. Here are three commonly used design patterns in Java:
    1. Singleton Pattern:
    - Purpose: This pattern ensures that a class has only one instance
and provides a global point of access to that instance. It's particularly
useful when you want to control access to shared resources, such as
configuration managers or connection pools.
    - Importance: The Singleton pattern is important because it provides
a controlled mechanism to prevent creating multiple instances of a class,
which can save memory and synchronize access to resources. Moreover,
it ensures that operations using the singleton are using the same instance
and data, which is critical in scenarios where consistent state or
coordinated actions are required.
    2. Observer Pattern:
    - Purpose: This pattern defines a one-to-many dependency between
objects. When one object (known as the subject or observable) changes
its state, all of its dependents (observers) are notified and updated
automatically. It's the underlying principle behind the
Model-View-Controller (MVC) architecture, where the model is the subject
and the views are observers.
    - Importance: The Observer pattern is vital for building systems
that are modular and decoupled. Changes in one part of the system don't
directly affect others, allowing for easier maintenance and scalability.
This pattern also makes it easier to add new observers without modifying
existing code, enhancing the system's extensibility.
    3. Factory Pattern:
    - Purpose: The Factory pattern provides an interface for creating
```

instances of a class, but it's up to the derived subclasses to implement the method and determine which class to instantiate. It promotes the loose coupling by eliminating the need to bind application-specific classes into the code. Instead, the code interacts only with the resulting interface or abstract class, so the system can evolve without altering the calling code.

- Importance: Using the Factory pattern, a system can be more flexible and scalable. If there's a need to introduce a new product or modify an existing one, it can be done without disturbing the existing code. It encapsulates the object creation process and separates it from the main logic of the program, promoting the Single Responsibility Principle. This ensures that a class is only responsible for one reason to change.

All these patterns, and many more, provide a vocabulary for developers, making it easier to communicate complex software designs and concepts. They embody solutions to recurring problems and can significantly speed up the development process, reduce errors, and improve code quality.

中文注释

当然！设计模式是针对软件设计中出现的常见问题的可重复使用的解决方案。它们是经验丰富的开发人员发现并经过长期改进的最佳实践。由于Java语言的面向对象特性，Java开发人员经常使用设计模式。下面是Java中常用的三种设计模式。

1. 单例模式

— 目的：该模式确保一个类只有一个实例，并提供一个访问该实例的全局点。若想控制对共享资源(如配置管理器或连接池)的访问，它尤为有用。

— 重要性：单例模式之所以重要，是因为它提供了一种受控机制，可防止创建一个类的多个实例，从而节省内存并同步访问资源。此外，它能确保使用单例的操作使用的是相同的实例和数据，这在需要一致状态或协调操作的场景中至关重要。

2. 观察者模式

— 目的：该模式定义了对象之间一对多的依赖关系。当一个对象(称为主体或可观察对象)改变其状态时，它的所有从属对象(观察者)都会收到通知并自动更新。这是模型—视图—控制器(MVC)架构的基本原理，其中模型是主体，视图是观察者。

— 重要性: 观察者模式对于构建模块化和解耦的系统至关重要。系统中某一部分的变化不会直接影响其他部分，这使得维护和伸缩变得更加容易。这种模式还能在不修改现有代码的情况下轻易地添加新的观察者，从而增强系统的可伸缩性。

3. 工厂模式

— 目的: 工厂模式提供了一个创建类实例的接口，但由派生的子类来实现方法并决定实例化哪个类。该模式下，不必将特定于应用程序的类绑定到代码中，从而促进松散耦合。相反，代码只与生成的接口或抽象类交互，因此系统可以在不改变调用代码的情况下不断发展。

— 重要性: 使用工厂模式，系统可以更具灵活性和可伸缩性。如果需要引入新产品或修改现有产品，可以在不影响现有代码的情况下完成。工厂模式封装了对象的创建过程，并将其与程序的主逻辑分离，从而促进单一责任原则(Single Responsibility Principle)，确保一个类只负责一个要更改的原因。

所有这些模式，以及更多其他的模式，都为开发人员提供了一个词汇表，使得复杂的软件设计和概念之间的交互变得更加容易。它们包含了反复出现的问题的解决方案，可以显著加快开发速度，减少错误，并提高代码质量。

很不错吧？ChatGPT 不仅为 Java 开发人员列出了常用的设计模式，还合理解释了这些设计模式的重要性，而这一切都是用流畅的自然语言完成的。

那么，这对 Java 开发人员究竟意味着什么呢？这其实是件意义重大的事情，这是因为现在开发人员可创建出能够“理解”自然语言文本的应用程序了。在 ChatGPT 等人工智能工具出现之前，开发人员通常需要使用正则表达式(也称为“regex”)在文本中执行基本的字符和字符串模式匹配。然而，这样的模式匹配与真正的自然语言理解相去甚远。现在，有了 ChatGPT 这样的工具，开发人员能够更深入地处理和理解文本内容，这无疑为 Java 开发领域带来了全新的可能性。

1.5 Regex 与 ChatGPT：对抗！

注意：

如果你对正则表达式的局限性，特别是它在自然语言理解和情感分析方面的无能为力已经有所了解，那么可以选择跳过接下来的内容。

我坚信，每个程序员在其职业生涯的某个阶段，都会遇到一些在正则表达式方面颇有造诣的人。正则表达式确实是一个强大的工具，它能够解析大量的文本，帮助程序员在代码中识别并提取出特定的文本模式。

然而，正则表达式也存在一个显著缺点，那就是它的可读性差。一旦编写完成，正则表达式往往难以被他人(甚至是最初编写它的开发人员)理解。

因此，我们有必要比较一下正则表达式与具备自然语言处理(NLP)和自然语言理解(NLU)功能的 ChatGPT 之间的差异。

代码清单 1-3 描述了一个虽然有些夸张但富有启发性的故事。它告诉我们，尽管正则表达式在查找文本中的单词和短语方面表现出色，却无法提供任何形式的自然语言理解。而这正是 ChatGPT 等 NLP 工具能为我们带来的价值。

代码清单 1-3　Sadstory.txt

```
    In the city of Buttersville,USA on Milkmaid street, there's a group
of three friends: Marion Yogurt, Janelle de Queso, and Steve Cheeseworth
III. On a hot summer's day, they heard the music from an ice cream truck,
and decided to buy something to eat.
    Marion likes strawberries, Janelle prefers chocolate, and Steve is
lactose intolerant. That day, only two kids ate ice cream, and one of
them bought a bottle of room-temperature water. The ice cream truck was
fully stocked with the typical flavors of ice cream.
```

中文注释——一个关于不吃冰淇淋的孩子的悲伤故事

在美国巴特尔斯维尔市的 Milkmaid 街，住着三个好朋友：Marion Yogurt、Janelle de Queso 和 Steve Cheeseworth III。在一个炎热的夏日，他们被冰淇淋车的音乐声所吸引，决定去买些冷饮消暑。

Marion 偏爱草莓口味，Janelle 偏爱巧克力口味，而 Steve 有乳糖不耐症。那天，只有两个孩子享用了冰淇淋，其中一个还顺便买了瓶常温水。车上的冰淇淋琳琅满目，令人垂涎欲滴。

1.5.1　分析问题 1：谁没有得到冰淇淋，为什么？

现在，我们来深入分析这个问题，并提出一些疑问。首先，我们要问的是：谁没有吃到冰淇淋，以及背后的原因是什么？答案显而易见，Steve 因为乳糖不耐症而没有吃到冰淇淋。然而，值得注意的是，故事中并没有明确提及 Steve 没有买冰淇淋，这就给正则表达式带来了匹配上的困难。

正则表达式擅长于查找文本中的关键字，如“没有”“没有冰淇淋”或孩子们的名字。但它只能基于这些模式是否出现来提供响应。例如，如果正则表达式将“没有”或“没有冰淇淋”与 Steve 的名字相匹配，就能显示出对应的文本模式。然而，正则表达式却无法解释为何 Steve 是那个没有冰淇淋的人，更无法提供任何与上下文相关的推理。

接下来，我们尝试将同样的故事提供给 ChatGPT，并提出问题：“谁没有吃到冰淇淋？”代码清单 1-4 将我们的问题与前面的故事整合在一起，作为 ChatGPT 的输入提示。

代码清单 1-4　提示。以悲伤故事作为提示

```
    Using the information in the following story, who didn't get any
ice cream and why?
    ###
    In the city of Buttersville,USA on Milkmaid street, there's a group
of three friends: Marion Yogurt, Janelle de Queso, and Steve Cheeseworth
III. On a hot summer's day, they heard the music from an ice cream truck,
and decided to buy something to eat.
    Marion likes strawberries, Janelle prefers chocolate, and Steve is
lactose intolerant. That day, only two kids ate ice cream, and one of
them bought a bottle of room-temperature water. The ice cream truck was
fully stocked with the typical flavors of ice cream.
```

中文注释

根据以下故事中的信息，我们要找出谁没有得到冰淇淋以及背后的原因是什么。

###

在美国巴特尔斯维尔市的 Milkmaid 街，住着三个好朋友：Marion Yogurt、Janelle de Queso 和 Steve Cheeseworth Ⅲ。在一个炎热的夏日，他们被冰淇淋车的音乐声所吸引，决定去买些冷饮消暑。

Marion 偏爱草莓口味，Janelle 偏爱巧克力口味，而 Steve 有乳糖不耐症。那天，只有两个孩子享用了冰淇淋，其中一个还顺便买了瓶常温水。车上的冰淇淋琳琅满目，令人垂涎欲滴。

在构建包含指令和数据的提示时，如上述内容所示，采用某种形式的分隔(如本例中的“###”)是一种推荐的做法。然而，在后续使用 Playground 或通过 Java 程序调用 ChatGPT API 时，你会发现更加高效且清晰的分隔方法。

因此，在向 ChatGPT 发出提示后，它会基于故事信息给出一个答案，如代码清单 1-5 所示。

代码清单 1-5　响应。ChatGPT 对分析问题 1 的响应

```
    Based on the given information, Steve is lactose intolerant and
therefore cannot eat ice cream. Therefore, Steve is the one who didn't
get any ice cream.
```

中文注释

根据给定的信息，Steve 由于乳糖不耐症而无法享用冰淇淋。因此，Steve 是没有吃到冰淇淋的人。

正如你所看到的，ChatGPT 可以利用 NLP 和 NLU(即人工智能)理解场景的上下文。它能够解析孩子们之间的关系、他们各自的喜好和 Steve 的乳糖不耐症。此外，ChatGPT 还能区分故事中的细节，如孩子们的姓氏、街道名称和城市名称，尽管它们都与解答问题无关，但都是与乳制品相关的有趣命名。

1.5.2　分析问题 2：哪个孩子可能会伤心？

现在，为了进一步证实正则表达式无法提供任何类型的 NLP 或 NLU，让我们使用一个新的术语——**情感分析**。冰淇淋车开走后，哪个孩子会伤心呢？

由于故事中没有提到任何孩子的感受或情绪，因此没有任何文本模式可以让任何正则表达式返回匹配结果。

但是，如果你向 ChatGPT 提出同样的问题，它将返回代码清单 1-6 所示的响应。

代码清单 1-6　响应。ChatGPT 对分析问题 2 的响应

```
    Since Steve is lactose intolerant and cannot eat ice cream, he would
be the kid left sad because he couldn't enjoy the ice cream like Marion
and Janelle.
```

中文注释

因为 Steve 有乳糖不耐症，不能吃冰淇淋，所以他会是那个因为不能像 Marion 和 Janelle 一样享用冰淇淋而伤心的孩子。

综上，ChatGPT 能够理解情景，通过信息进行推理，提供正确的答案并给出解释。

1.6　了解更多有关 ChatGPT API 的信息需要先了解一些术语

首先，在开始使用 ChatGPT 和 OpenAI API 之前，你应该先熟悉一些词语和术语。因此，在以编程方式使用 ChatGPT 时，要先弄清楚“模型”“提示”“词元”和“温度”的定义。

1.6.1 模型

作为一名Java开发人员，一听到“模型”这个词，你或许会立刻联想到面向对象编程和 Java 类中对于现实世界实体的表示，如“对象模型”。同样，如果你有过数据库的使用经验，那么“模型”这个词也可能会让你想到数据库中数据及其关系的表示，如“数据模型”。

然而，在使用 ChatGPT API(以及一般的人工智能技术)时，你需要暂时放下这两个定义，因为它们在这里并不适用。在人工智能领域，“模型”指的是预训练的神经网络。

请记住，正如我之前提到的，阅读本书并不需要具备机器学习方面的博士学位。那么，什么是神经网络呢？简单来说，神经网络是人工智能系统的核心组件，其设计目的是通过模拟人脑的工作方式，利用相互连接的人工神经元层来处理和分析数据，如图 1-1 所示。这些网络可在大量数据的基础上进行训练，以学习模式、发现关系并进行预测。

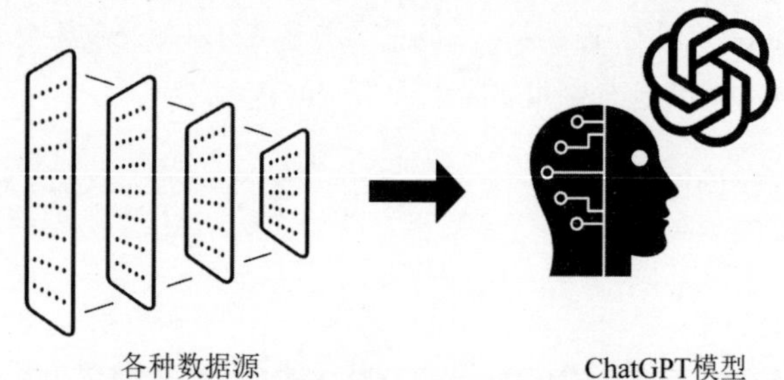

图 1-1　在大量数据上训练的人工智能模型

在人工智能领域，“预训练模型”指的是在提供给开发人员使用之前，已经在特定任务或数据集上训练过的神经网络。这一训练过程包括让模型接触大量标注和分类(也称为“注释”)的数据，并调整其内部参数，以优化其在特定任务中的表现。

表 1-1 列出 OpenAI 为开发人员提供的一些模型，以帮助应用程序实现 AI 功能。

表 1-1　OpenAI 为开发人员提供的一些模型

模型名称	说明
GPT-4	GPT-4 是 OpenAI 最新一代的 GPT 模型集。GPT 是 Generative Pre-trained Transformer 的缩写，这些模型经过训练可以理解自然语言和多种编程语言。GPT-4 模型将文本和图像作为提示输入，并提供文本作为输出。 现有的 GPT-4 模型包括： • gpt-4 • gpt-4-32k • gpt-4-vision
GPT-3.5	GPT-3.x 是 OpenAI 的 GPT 模型集的上一代产品。最初于 2022 年 11 月对外发布的 ChatGPT 使用的是 GPT 3。 可用的 GPT-3 模型包括： • gpt-3.5-turbo • gpt-3.5-turbo-16k
DALL-E	DALL-E 模型可根据自然语言提示生成和编辑图像。 在第 4 章中，我们将使用 DALL-E 模型来可视化你最喜欢的播客节目中正在讨论的对话。 以下是一些可用的 DALL-E 模型： • dall-e-3 • dall-e-2
TTS	TTS 模型将文本转换成音频，效果出奇地好。大多数情况下，音频几乎与人声无异。 一些可用的 TTS 模型包括： • tts-1 • tts-1-hd
Whisper	简单地说，Whisper 模型就是将音频转换成文本。 在本书中，我们将使用 Whisper 模型搜索播客中的文本

(续表)

模型名称	说明
嵌入	嵌入模型可将大量文本转换为文本中字符串之间关系的数字表示。这有什么用呢？嵌入允许开发人员使用自定义数据集完成特定任务。如此，这意味着可在与应用程序相关的特定数据上训练嵌入模型。这样，就可执行以下操作： • 在文本正文中搜索 • 对数据进行聚类，从而根据相似度对文本字符串进行分组 • 获取推荐(推荐具有相关文本字符串的项) • 检测异常(识别关联度低的异常值) • 测量多样性(分析相似性分布) • 数据分类(根据最相似的标签对文本串进行分类)
调节	调节模型是可以检测文本是否敏感或不安全的微调模型。这些模型可以分析文本内容，并按照以下类别进行分类： • 仇恨 • 仇恨/威胁 • 骚扰 • 骚扰/威胁 • 自残 • 自残/意图 • 自残/指令 • 性行为 • 性/未成年人 • 暴力 • 暴力/图形 可用的调节模型有 • Text-moderation-latest • Text-moderation-stable
过时和弃用	自 ChatGPT 发布以来，OpenAI 一直在继续支持其旧版人工智能模型，但这些模型已被标记为“过时”或“弃用”。它们仍然存在，但其他更准确、更快速、更便宜的模型也已发布

注意：

这绝不是 OpenAI 为开发人员提供的可用模型的详尽列表！随着新模型的发布，旧模型将被标记为过时模型或弃用模型。因此，请务必查看 OpenAI 文档模型列表中的可用模型列表，了解最新信息：https://platform.openai.com/docs/models。

1.6.2　当谈论词元时，指的是 StringTokenizer 而非 Access Token

在使用第三方 API 时，你可能认为词元与访问词元的意义相同，后者通常是一个 UUID，可以让你识别自己的身份并与服务保持对话。好了，现在先不考虑这个定义。

现在，作为一名 Java 开发人员，你可能已经有机会使用 java.util.StringTokenizer 这个类来获取一个字符串，并将其拆分成一个由较小字符串组成的数组，这样你就可以根据需要遍历字符串。例如，如果你有一段文本，可将分隔符设为“.”，这样就可得到段落中句子的数组。

好消息是，OpenAI API 的“词元”概念与 Java 的概念非常相似，都是文本片段。对于 OpenAI API 而言，词元就是长度约为 4 个字符的文本片段。仅此而已，没有其他特别之处。

那么，如果词元是大约 4 个字符的文本块，我们为什么还需要关心它？

在使用 OpenAI 文本模型时，开发人员需要注意词元限制，因为它们会影响 API 调用的成本和性能。例如，gpt-3.5-turbo 模型的词元限制为 4096 个词元，而 gpt-4-vision 模型的词元限制为 128 000 个(大约相当于一本 300 页小说的大小)。模型的词元限制称为上下文窗口。

因此，开发人员需要考虑作为模型输入的提示和输出的长度，确保它们符合模型的词元限制。

表 1-2 列出一些具有词元限制的最新模型及其价格。

表 1-2 模型、词元限制和每个词元的成本

模型	最大词元数	词元输入成本	词元输出成本
gpt-4	8 192	$0.03/1K 词元	$0.06/1K 词元
gpt-4-32k	32 768	$0.06/1K 词元	$0.12/1K 词元
gpt-4-vision	128 000	$0.01/1K 词元	$0.03/1K 词元
gpt-3.5-turbo-instruct	4 096	$0.0015/1K 词元	$0.002/1K 词元
gpt-3.5-turbo-16k	16 384	$0.0010/1K 词元	$0.002/1K 词元
text-embedding-ada-002	8192	$0.0001/1K 词元	

1.6.3 温度关乎创造力

当然，ChatGPT 没有生命，所以它无法像我们人类一样思考。不过，通过调整 ChatGPT API 提示中的“温度”设置，可以让其响应更具创造性，如图 1-2 所示。但是，如果想充分利用它的潜力，了解它能理解什么是至关重要的。

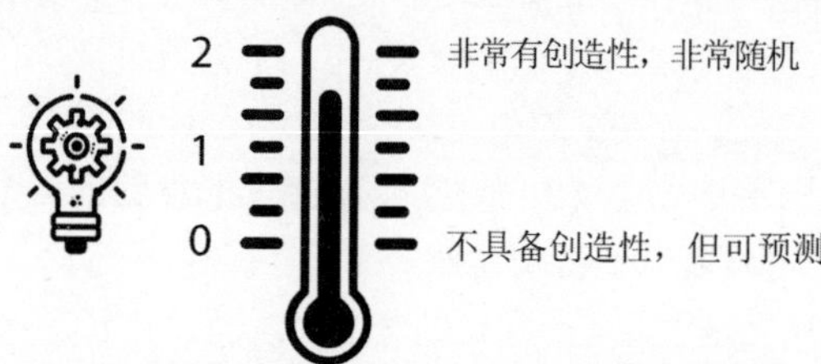

图 1-2 修改温度以获得更多或更少的创造性反应

1.7 OpenAI Playground 入门

现在，是时候将我们迄今为止学到的概念付诸实践了！不过，首先需要一个 OpenAI 的开发者账户，并创建一个 API 密钥。

请访问以下网址创建开发者账户和 API 密钥：https://platform.openai.com/account/api-keys。

如图 1-3 所示，可随意为 API 密钥命名。

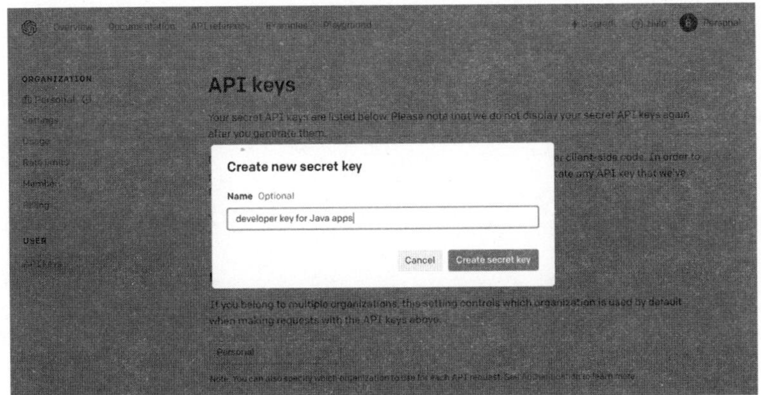

图 1-3　在访问 Playground 或进行 API 调用之前，需要有一个 API 密钥

需要注意，创建 API 密钥时，需要向 OpenAI 提供一张信用卡，以便 OpenAI 向你收取使用其模型的费用。

现在你已经获得了 API 密钥，让我们直接进入 Chat Playground，网址如下：https://platform.openai.com/playground。

进入 Playground 后，单击顶部的组合框，选择 Chat 选项以结束 Chat Playground，如图 1-4 所示。

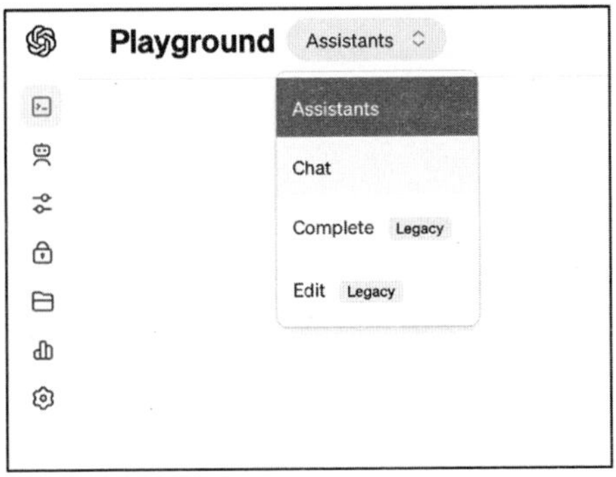

图 1-4　进入 Playground 后，选择 Chat 选项

图 1-5 为 Chat Playground 界面，其中某些部分已编号，以方便识别。

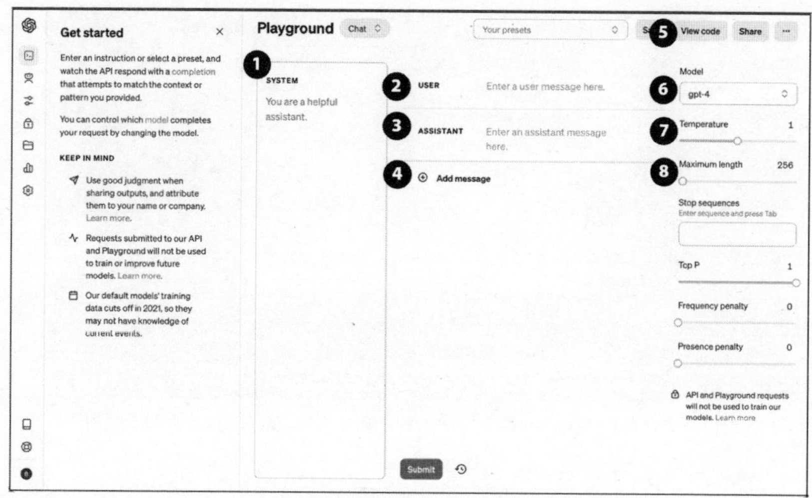

图 1-5 Chat Playground 乍一看可能有点令人费解

1. 系统

如你所见，Chat Playground 的用户界面要比其他人使用的 ChatGPT 网站复杂得多。因此，先谈谈系统(SYSTEM)字段(见图 1-5 第❶项)。

个人看来，ChatGPT 可以说是“一种强大的人工智能……具有健忘症”。因此，当你以编程方式使用 ChatGPT 时，需要告知系统在对话中的身份！

如图 1-6 所示，你可以一窥 ChatGPT 在对话中可以扮演的数千种不同角色。

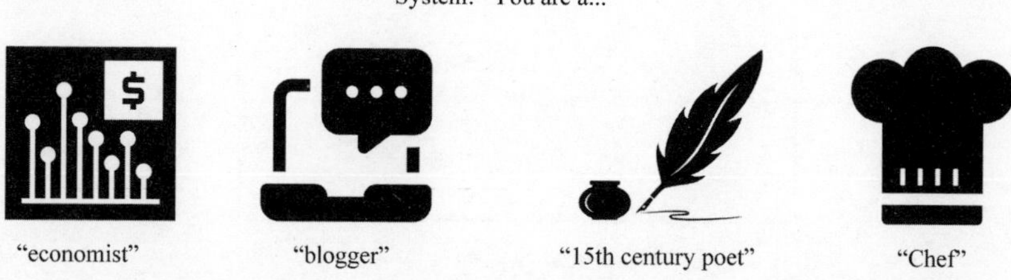

图 1-6 可通过 Chat Playground 中的系统字段设置 ChatGPT 在对话中扮演的角色

2. 用户

ChatGPT Playground 中的用户(USER)字段(图 1-5，第❷项)是向

ChatGPT 输入提示的地方，可以是任何你想要的内容，例如“描述远程医疗将如何影响医疗行业”。

3. 助手(可选)

初次加载 Chat Playground 时，无法看到助手(ASSISTANT)字段(图 1-5 第❸项)。若要让它出现，需要单击“添加信息”(Add message)旁的“+”符号。此刻，你可能会问自己:“为什么需要这个字段？”这个问题问得好。如果你想让 ChatGPT 记住它在之前的对话中已经告诉过你的内容，就需要在助手字段中输入它已经告诉过你的任何你认为相关的内容，以便继续对话。请记住，虽然它是一个非常强大的人工智能，但它有健忘症！

4. 添加信息(可选)

若需要在对话中添加一条助手信息或另一条用户信息，单击添加信息(Add message)旁的+符号(图 1-5，第❹项)。现在，你可能又会问：“既然我可以在上面原来的用户字段中输入我想要的内容，那么在对话中添加另一条用户信息又有什么意义呢？”问得好。

如果你想把命令和数据分开，那么可以使用单独的用户信息。

还记得在本章前面的代码清单 1-4 中，我们必须使用“###”将 ChatGPT 的命令与其需分析的数据分开吗？现在不需要了，因为命令是第一条用户信息，而数据是第二条。

5. 查看代码(可选)

使用 Playground 提交提示后，你可以单击“查看代码”(View code)按钮(图 1-5，第❺项)，以便查看使用任何支持的语言发送相同提示所需的代码。

你可能注意到 Java 并不是官方支持的语言，但我们将在第 2 章中解决这个问题，届时将使用 ChatGPT 作为结对程序员，并自行将其 REST 接口移植到 Java 中。

6. 模型(可选)

在本章前面，我们谈到了可供开发者使用的各种模型。单击模型(Model)字段可查看可用模型列表。

你可能还会看到，某些模型的名称中包含了月份和日期，这只是该模型的简要说明。通过程序选择一个说明，开发人员可以在某种程度上预测从 ChatGPT 收到的响应，因为当前模型一直在更新。

7. 温度(可选)

如本章前面所述，温度选择器的范围在 0 到 2 之间，便可选择响应的“随机性”。

8. 最大长度(可选)

还记得本章前面关于词元的讨论吗？通过在此最大长度内任意选择，可调整响应中的词元数(直接影响单词数)。

1.8 立即尝试！试用“系统”角色

现在我们已经熟悉了 Chat Playground 的一些特性，让我们使用上面讨论的设置发送第一条提示信息。代码清单 1-7 和代码清单 1-8 使用了相同的提示，要求 ChatGPT 提供几段关于远程医疗的内容，但系统的角色却大不相同。

代码清单 1-7 提示。作为一名研究人员讨论远程医疗的利与弊

```
System: You are a strictly factual researcher
User: Write 3 paragraphs on pros and cons of telemedicine
```

中文注释

系统：你是一名严格的事实研究员

用户：就远程医疗的利弊写 3 段文字

代码清单 1-8 提示。作为一名极有主见的健康博主讨论远程医疗的利与弊

```
System: You are a highly opinionated health blogger who always has stories with first hand experience
User: Write 3 paragraphs on pros and cons of telemedicine
```

中文注释

系统：你是一位极有主见的健康博主，总是有亲身经历的故事。

用户：就远程医疗的利弊写 3 个段落

我们鼓励你自己尝试这两个提示，看看有什么响应。调整温度和词元长度的设置，以熟悉这些参数对结果的影响。

1.9　小结

本章进一步讲述开发者如何使用 ChatGPT。介绍了 Chat Playground 的一些基础知识，这是一个供开发者与 ChatGPT API 交互的 Web 界面。

讨论了如何在 Chat Playground 中设置系统、用户和助手角色，以及如何调整温度和最大输出长度等设置。

介绍了使用 Chat Playground 所需的一些参数和术语，如模型、温度和词元。熟悉 Chat Playground 的参数对于了解如何使用 REST API 至关重要，因为 Chat Playground 是 REST API 所提供功能的一个子集。

在第 2 章中，将讨论如何使用 ChatGPT 作为“结对程序员”，并将官方支持的 ChatGPT REST 接口移植到 Java 中。

第2章

使用 ChatGPT 作为 Java 结对程序员

我非常喜欢 XP(极限编程)的一些实践，尤其是结对编程。无论你喜欢哪种结对编程，都需要两个工程师坐在同一个屏幕前，共同解决同一个问题。这样做的最大好处之一是可以用全新的视角来看待问题，当然，现在有两个工程师“接触”过代码库，而不是一个人。有时，你可以让一名工程师编写代码，另一名工程师编写测试和注释。无论怎样分工，合作总是好的。

现在，用于 ChatGPT 及其他模型的 OpenAI REST API 已受到 Python、Typescript 和 cURL 的正式支持。

还有一些 Java API 是由第三方开发人员创建的，但最大的问题(在我看来)是这个领域正在迅速变化。OpenAI 正在不断更新其模型和 HTTP 接口，因此，他们会经常添加或弃用一些特性和功能。如果你选择在项目中使用第三方 Java API，你将面临使用过时或与 OpenAI REST API 不同步的 API 的风险。

因此，在本章中，将使用 ChatGPT 作为结对程序员，并简单地将官方 OpenAI REST API 直接移植到 Java 中。每当 OpenAI 对其官方支持的语言和接口进行任何修改时，就可以立即更新自己的库。开始吧！

2.1 创建第一个 Java ChatGPT 应用程序：ListModels.java

实际上，我们要同时完成两项任务：使用 OpenAI API 创建一个基本的 Java 应用程序，并在此过程中验证我们是否正确获取了 API 密钥。因此，如果你还没有这样做，请按照第 1 章中的说明创建 OpenAI 开发者账户并获取 API 密钥。接下来，本书中的所有代码示例都需要有效的 API 密钥。

2.2 列表模型端点

可以调用的最基本(但也必不可少)的服务之一是列表模型端点。你可能会问为什么？通过列表模型端点，可通过 REST API 获取当前可供开发人员使用的所有 AI 模型。

2.2.1 创建请求

表 2-1 列出了调用列表模型端点需要的所有 HTTP 参数。

表 2-1 调用列表模型端点所需的 HTTP 参数

HTTP 参数	说明
Endpoint URL(端点 URL)	https://api.openai.com/v1/models
Method(方法)	GET
Header(标题)	Authorization：Bearer $OPENAI_API_KEY

2.2.2 处理 JSON 响应

成功调用列表模型端点后，服务将提供一个 JSON 响应，其结构如表 2-2 所示。

表 2-2　模型 JSON 对象的结构

字段	类型	描述
object	String	总是返回 list(列表)
data	Array	OpenAI 提供的 AI 模型列表
↳id	String	AI 模型的唯一 ID，即模型的全称
↳object	String	总是返回 model(模型)
↳created	integer	模型的创建日期
↳owned_by	String	拥有模型的组织名称
permission	Array	用于模型权限的元数据数组
root	String	这通常是 data.id 字段的副本，也就是模型的名称
parent	String	目前是一个未使用的字段

2.2.3　模型(JSON 格式)

注意：

由于 JSON 对象可以包含数组(很难在表格中表示)，我们使用符号"↳"来表示数组的元素。从上表可以看出，id、object、created 和 owned_by 都是 JSON 响应中 data 数组的元素。在本书后面的表格中，也能看到该符号。

现在我们已经掌握了 HTTP 请求和响应的详细信息，可使用代码清单 2-1 中的提示创建 ListModel.java 应用程序。

代码清单 2-1　提示。使用 ChatGPT 创建 ListModels.java 应用程序

```
System: You are a Java developer
User: Take the following cURL command and convert it to Java.
Name the file, "ListModels.java"
User: curl https://api.openai.com/v1/models \
  -H "Authorization: Bearer $OPENAI_API_KEY"
```

中文注释

系统：你是一名 Java 开发人员

用户：执行以下 cURL 命令并将其转换为 Java 格式。将文件命名为 ListModels.java

用户：curl https://api.openai.com/v1/models \ -H "Authorization: Bearer $OPENAI_API_KEY"

发送提示后，ChatGPT 会提供如代码清单 2-2 所示的响应。

代码清单 2-2 响应。ListModels.java

```
import java.io.BufferedReader;
import java.io.InputStreamReader;
import java.net.HttpURLConnection;
import java.net.URL;

public class ListModels {

  public static void main(String[] args) {
  try {
    // Create URL object for the API endpoint
    URL url = new URL("https://api.openai.com/v1/
    models");

    // Create HttpURLConnection object
    HttpURLConnection conn = (HttpURLConnection) url.
    openConnection();

    // Set request method
    conn.setRequestMethod("GET");

    // Set Authorization header
    conn.setRequestProperty("Authorization", "Bearer
    $OPENAI_API_KEY");

    // Get response code
    int responseCode = conn.getResponseCode();

    // If the request was successful (status code 200)
    if (responseCode == HttpURLConnection.HTTP_OK) {
      // Read the response from the API
      BufferedReader in = new BufferedReader(new
      InputStreamReader(conn.getInputStream()));
      String inputLine;
      StringBuilder response = new StringBuilder();
```

```
        while ((inputLine = in.readLine()) != null) {
          response.append(inputLine);
        }
        in.close();

        // Print the response
        System.out.println(response.toString());
      } else {
        System.out.println("Request failed. Response
        Code: " + responseCode);
      }

      // Close the connection
      conn.disconnect();
    } catch (Exception e) {
      e.printStackTrace();
    }
  }
}
```

太棒了！现在，只需要将$OPENAI_API_KEY 替换为实际 API 密钥的值即可。

运行该类后，将得到一个 JSON 文件，其中包含 OpenAI API 提供的所有模型，如代码清单 2-3 所示。

代码清单 2-3 响应。运行 ListModels.java 后的部分 JSON 响应

```
{
  "object": "list",
  "data": [
  {
    "id": "text-search-babbage-doc-001",
    "object": "model",
    "created": 1651172509,
    "owned_by": "openai-dev"
  },
  {
    "id": "gpt-4",
    "object": "model",
    "created": 1687882411,
```

```
    "owned_by": "openai"
  },
  {
    "id": "gpt-3.5-turbo-16k",
    "object": "model",
    "created": 1683758102,
    "owned_by": "openai-internal"
  },
  {
    "id": "curie-search-query",
    "object": "model",
    "created": 1651172509,
    "owned_by": "openai-dev"
  },
  {
    "id": "text-davinci-003",
    "object": "model",
    "created": 1669599635,
    "owned_by": "openai-internal"
  },
  {
    "id": "text-search-babbage-query-001",
    "object": "model",
    "created": 1651172509,
    "owned_by": "openai-dev"
  },
  {
    "id": "babbage",
    "object": "model",
    "created": 1649358449,
    "owned_by": "openai"
  },
...
```


由于可供开发人员使用的模型数量庞大，代码清单 2-3 只列出了部分模型！不过好消息是，附录 A 中的表格提供了完整的响应内容。

既然我们可以通过编程获取可用模型列表，那么现在就可以使用 Java 向 ChatGPT 发送提示信息了，这可以通过聊天端点来实现。

2.3　聊天端点

聊天端点(以前称为“聊天完成”)是一种 REST 服务，基本上是你在 Chat Playground(聊天平台)中可以做的事情的一对一表示；因此，这项服务对你来说简直就是家常便饭。

2.3.1　创建请求

表 2-3 列出了调用聊天端点需要的所有 HTTP 参数。

表 2-3　聊天端点的 HTTP 参数

HTTP 参数	说明
Endpoint URL(端点 URL)	https://api.openai.com/v1/chat/completions
Method(方法)	POST
Header(标题)	Authorization：Bearer $OPENAI_API_KEY
Content-Type(内容类型)	application/json

表 2-4 呈现了聊天端点请求体所需的 JSON 对象格式。快速浏览后可以发现，要成功调用服务，实际上只需要几个字段。

表 2-4　聊天 JSON 对象的结构

字段	类型	是否必需	描述
model	String	必需	用于聊天完成的模型 ID 兼容的模型包括 • gpt-4 • gpt-4-0613 • gpt-4-32k • gpt-4-32k-0613 • GPT-3.5-turbo • GPT-3.5- turbo -0613 • GPT-3.5- turbo -16K • GPT-3.5- turbo -16K-0613

(续表)

字段	类型	是否必需	描述
messages	Array	必需	正在进行的对话中的消息数组。 数组中的每条信息都有两个属性：role(角色)和 content(内容)
↳role	String	必需	指定信息的角色，可以是以下任何一种： • system(系统) • user(用户) • assistant(助手) • tool(工具)
↳content	String	必需	包含指定角色的信息文本
tools	Array	可选	允许你指定模型可以调用的工具列表。目前，唯一支持的工具类型是函数。 通过该参数，你可以定义一组函数；模型为这些函数生成 JSON 输入
↳type	String	必需	这是工具的类型，可以是 • function(函数)
↳function	Array	可选	模型可在“聊天完成”时调用的函数数组
↳↳name	String	必需	要调用的函数名称。 有效名称必须是 a～z、A～Z、0～9，或包含下画线和破折号。 最大长度为 64 个字符
↳↳description	String	可选	对函数作用的描述。 这有助于模型决定是否在聊天完成中调用该函数
↳↳parameters	JSON 对象	必需	函数接收的 JSON 模式对象格式的参数

(续表)

字段	类型	是否必需	描述
tool_choice	String 或 JSON 对象。默认值：当请求中不包含任何函数时为 none，当请求中包含函数时为 auto	可选	这允许你决定模型应调用哪个函数(如果有的话)。 设置为 none(无)时，模型将不调用任何函数，只生成消息响应。 当设置为 auto(自动)时，模型可根据其内部决策过程，灵活选择生成消息响应或调用函数
temperature	Number 或 null。默认值：1	可选	有效值范围在 0 到 2 之间。 控制模型输出的随机性。 最佳做法是调整 top_p 或 temperature，但不能同时调整两者
top_p	Number 或 null。默认值：1	可选	有效值范围在 0 和 1 之间。 表示是考虑少数可能性(0)还是所有可能性(1)。 最佳做法是调整 top_p 或 temperature，而不是同时调整两者
n	Integer 或 null。默认值：1	可选	指定模型应为每个输入信息生成多少个聊天完成选项
stream	Boolean 或 null。默认值：false	可选	如果设置为 true，部分信息更新将作为服务器发送的事件发送。 这意味着词元在可用时将作为纯数据事件发送，流将以“data: [DONE]”结束
stop	String、Array 或 null。默认值：null	可选	最多可以提供 4 个序列，以便 API 停止生成更多词元。 这对于控制响应的长度或内容非常有用
max_tokens	Integer 或 null。默认值：inf	可选	该参数设置了生成的聊天完成(Chat Completion)中的词元的最大数量
response_format	JSON 对象	可选	有两个选项：{"type":"json_object"}，用于 JSON 对象响应；{"type":"text"}，用于文本响应

(续表)

字段	类型	是否必需	描述
seed	Integer 或 null	可选	通过指定种子，系统将尝试生成可重复的结果。 理论上，这意味着如果使用相同的种子值和参数重复请求，就会得到相同的结果。 为在后续请求中获得种子值，可复制上次响应中的 system_fingerprint
presence_penalty	Number 或 null。 默认值：0	可选	介于-2.0 和 2.0 的数字。 正值会根据新词元是否出现在对话历史中对其进行惩罚，从而鼓励模型谈论新话题
frequency_penalty	Number 或 null。 默认值：0	可选	介于-2.0 和 2.0 的数字。 正值会根据词元在对话历史中的现有频率对其进行惩罚，从而降低逐字重复相同行的可能性
logit_bias	JSON 映射。 默认值：null	可选	允许修改特定词元在完成时出现的可能性。 可以提供一个 JSON 对象，将词元(由标记化器中的标记 ID 指定)映射到相关的偏置值(从-100 到 100)。 该偏置值会在采样前添加到模型的 logit 中
user	String	可选	这是一个唯一的 ID，你可以选择生成它来代表你的终端用户。 这将有助于 OpenAI 监控和检测滥用行为

注意：

在本书中，将把 stream(流)参数设置为默认值，即 false。这意味着将以单个 HTTP 响应的形式一次性接收来自 ChatGPT 的结果。

不过，在某些情况下，你会希望将此设置为 true。比如说，你正在构建一个交互式语音聊天机器人。又比如说，你希望将 ChatGPT 中的文本转换为音频，这样用户就能听到声音响应。这种情况下，你肯定希望将 stream

参数设置为 true。这是为什么呢？因为当响应被流式传输回 Java 应用程序时，你就有机会将文本片段转换为音频。这样，你就可以在接收更多文本的同时，将文本片段转换为音频。这将使终端用户的响应看起来更自然，并让对话听起来像真正的对话。

2.3.2　聊天(JSON)

代码清单 2-4 是正确调用聊天端点的 JSON 对象的示例。

代码清单 2-4　聊天 JSON 对象示例

```
{
  "model": "gpt-3.5-turbo",
  "messages": [
    {
      "role": "system",
      "content": "You are a product marketer"
    },
    {
      "role": "user",
      "content": "Explain why Java is so widely used in the
      enterprise "
    }
  ],
  "temperature": 1,
  "max_tokens": 256,
  "top_p": 1,
  "frequency_penalty": 0,
  "presence_penalty": 0
}
```

2.3.3　处理响应

成功调用聊天端点后，API 会出现一个聊天完成对象(Chat completion)，如果启用了流处理，则会出现一个完成块流。下面是聊天完成对象的详细信息。

2.3.4 聊天完成(JSON)

表 2-5 总结了聊天完成 JSON 对象的结构。

表 2-5 聊天完成 JSON 对象的结构

字段	类型	描述
id	String	聊天完成的唯一标识符
object	String	总是返回文字“chat.completion”
system_fingerprint	String	如果你想重现之前对话的结果，请在后续请求中使用此参数作为“种子”
created	Integer	聊天完成的时间戳
model	String	聊天完成所使用的模型
choices	Array	可用的聊天完成选项列表。 如果在聊天 JSON 请求中使用参数 n 指定所需的响应数，就可以获得多个信息选项。请参见表 2-4
↳index	Integer	列表中选择的索引
↳message	Array	模型生成的聊天完成消息
↳finish_reason	String	每个响应都包含一个 finish_reason。finish_reason 的可能值如下。 stop：API 返回的完整信息，或通过 stop 参数提供的某个停止序列终止的信息。 length(长度)：由于请求中的 max_tokens 参数或模型本身的词元限制，模型输出不完整。 tool_call：模型调用了一个工具，如 function。 content_filter(内容过滤)：由于违反了内容过滤器，响应被终止。 null(空)：API 响应仍在进行中或未完成
usage	Array	完成请求的使用统计信息，包括提示和总请求数量
↳prompt_tokens	Integer	提示中使用的词元数量
↳completion_tokens	Integer	响应中使用的词元数量
↳total_tokens	Integer	请求和响应中所有词元的总和

代码清单 2-5 是调用 Chat Endpoint(聊天端点)后的 JSON 响应示例。

代码清单 2-5　聊天完成 JSON 对象

```
{
  "id": "chatcmpl-7wUOFQ3S34scDLmrLdWTTqvHmXztQ",
  "object": "chat.completion",
  "created": 1694174199,
  "model": "gpt-3.5-turbo-0613",
  "choices": [
    {
      "index": 0,
      "message": {
        "role": "assistant",
        "content": "Java is widely used in the enterprise because it is platform-independent, allowing applications to run on any system. Additionally, Java has a large and mature ecosystem with a vast array of libraries, frameworks, and tools, making it easier for developers to build robust and scalable enterprise applications."
      },
      "finish_reason": "stop"
    }
  ],
  "usage": {
    "prompt_tokens": 32,
    "completion_tokens": 55,
    "total_tokens": 87
  }
}
```


2.4　等等，我的提示中有多少词元

到了一定程度，你就会开始考虑计划发送到 ChatGPT 的提示，并对要使用的模型的词元限制(和成本)进行大量思考。如果你忘记了，请务必参阅表 1-2，查看模型和词元价格的列表。此外，OpenAI 还创建了一个简单

易用的网站，可以查看提示中的词元数量，如图 2-1 所示。

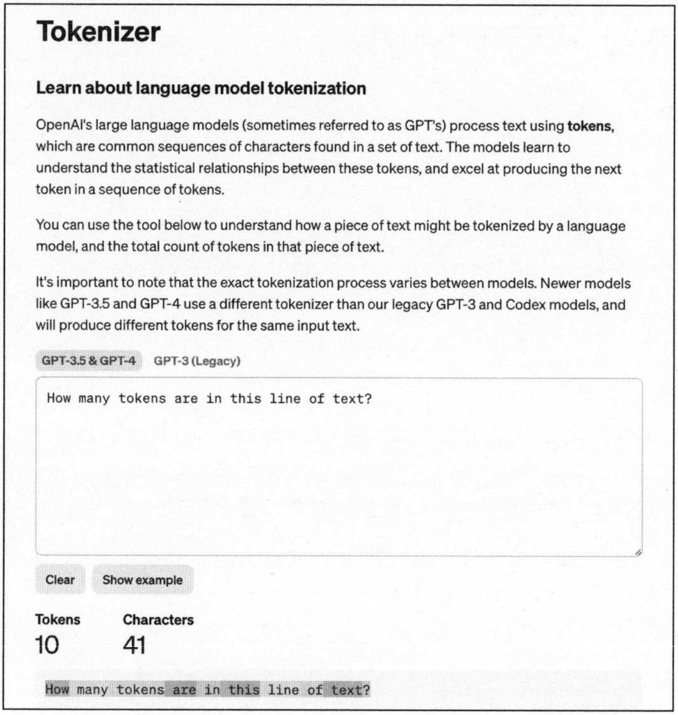

Tokenizer

Learn about language model tokenization

OpenAI's large language models (sometimes referred to as GPT's) process text using **tokens,** which are common sequences of characters found in a set of text. The models learn to understand the statistical relationships between these tokens, and excel at producing the next token in a sequence of tokens.

You can use the tool below to understand how a piece of text might be tokenized by a language model, and the total count of tokens in that piece of text.

It's important to note that the exact tokenization process varies between models. Newer models like GPT-3.5 and GPT-4 use a different tokenizer than our legacy GPT-3 and Codex models, and will produce different tokens for the same input text.

GPT-3.5 & GPT-4 GPT-3 (Legacy)

How many tokens are in this line of text?

Clear Show example

Tokens 10 Characters 41

How many tokens are in this line of text?

图 2-1 ChatGPT 词元生成器可以快速计算提示中的词元数量

ChatGPT 词元计数器

https://platform.openai.com/tokenizer

2.5 创建下一个 Java 应用程序 ChatGPTClient.java

现在是创建我们自己的 ChatGPTClient 的时候了。让我们直接开始，通过 ChatGPT Playground 向 ChatGPT 提供它需要的信息。

代码清单 2-6　创建 ChatGPTClient.java 的初始对话

```
System: You are a Java developer
User: Convert the following cURL command to Java. Make sure the
URL and API keys to the API are variables. I want to use the
Jackson library to create the JSON object in the request. Name
the main class, ChatGPTClient, and create helper classes if
necessary.

User:
curl https://api.openai.com/v1/chat/completions \
  -H "Content-Type: application/json" \
  -H "Authorization: Bearer $OPENAI_API_KEY" \
  -d '{
  "model": "gpt-3.5-turbo",
  "messages": [
    {
      "role": "system",
      "content": "You are a product marketer"
    },
    {
      "role": "user",
      "content": "Explain why Java is so widely used in the
      enterprise "
    }
  ],
  "temperature": 1,
  "max_tokens": 256,
  "top_p": 1,
  "frequency_penalty": 0,
  "presence_penalty": 0
}'
```

中文注释

系统：你是 Java 开发人员

用户：将以下 cURL 命令转换为 Java。确保 API 的 URL 和 API 密钥是变量。我想使用 Jackson 库在请求中创建 JSON 对象。命名主类 ChatGPTClient，必要时创建辅助类。

用户：(略)

ChatGPT 的响应是一个良好开端，但并没有完全按照我的要求提供一切回答。由于我没有明确指定HTTP调用的方式，ChatGPT 决定使用 Apache HTTPClient 库。因此，我添加了以下用户消息(代码清单 2-7)。

代码清单 2-7　提示。在会话中添加附加信息以创建 ChatGPTClient.java

```
User: I have a Java 11 environment, so use Java 11 HTTP classes
for the API calls. I'd rather use pure Java instead of Apache
HttpClient
```

中文注释

用户：我有一个 Java 11 环境，所以要使用 Java 11 HTTP 类来调用 API。我宁愿只使用 Java 而非 Apache HttpClient

这里的目标是进行对话。指望 ChatGPT 在一次提示中就能完全满足你的所有要求是不切合实际的。即使是开发人员也不会这么做！因此，请随时与 ChatGPT 进行对话，并向它说明它可能猜测或假设错误的地方。

代码清单 2-8 为 ChatGPTClient.java 的最终结果。

代码清单 2-8　响应。ChatGPTClient.java

```
import com.fasterxml.jackson.core.JsonProcessingException;
import com.fasterxml.jackson.databind.ObjectMapper;
import com.fasterxml.jackson.annotation.JsonProperty;

import java.io.BufferedReader;
import java.io.IOException;
import java.io.InputStreamReader;
import java.io.OutputStream;
import java.net.HttpURLConnection;
import java.net.URL;
import java.util.ArrayList;
import java.util.List;

public class ChatGPTClient {

    public static void main(String[] args) {
```

```
String openAIKey = "insert your API key here";
String endpoint = "https://api.openai.com/v1/chat/
completions";
String model = "gpt-3.5-turbo";
float temperature = 1.0f;
int max_tokens = 256;
float top_p = 1.0f;
int frequency_penalty = 0;
int presence_penalty = 0;

List<Message> messages = new ArrayList<>();
messages.add(new Message("system", "You are a product
marketer."));
messages.add(new Message("user", "Explain why Java is
so widely used in the enterprise"));

String jsonInput = null;
try {
   ObjectMapper mapper = new ObjectMapper();
   Chat chat = new Chat(model, messages, temperature,
   max_tokens, top_p, frequency_penalty, presence_
   penalty);
   jsonInput = mapper.writeValueAsString(chat);
} catch (JsonProcessingException e) {
   e.printStackTrace();
}

try {
   URL url = new URL(endpoint);
   HttpURLConnection connection = (HttpURLConnection)
   url.openConnection();
   connection.setRequestMethod("POST");
   connection.setRequestProperty("Content-Type",
   "application/json");
   connection.setRequestProperty("Authorization",
   "Bearer " + openAIKey);
   connection.setDoOutput(true);

   OutputStream outputStream = connection.
   getOutputStream();
   outputStream.write(jsonInput.getBytes());
   outputStream.flush();
   outputStream.close();

   int responseCode = connection.getResponseCode();
```

```
            if (responseCode == HttpURLConnection.HTTP_OK) {
            BufferedReader reader = new BufferedReader(new
            InputStreamReader(connection.
            getInputStream()));
            StringBuilder response = new StringBuilder();
            String line;
            while ((line = reader.readLine()) != null) {
            response.append(line);
         }
         reader.close();

         // Print the response
         System.out.println(response.toString());
         } else {
            System.out.println("Error: " + responseCode);
         }
         connection.disconnect();
      } catch (IOException e) {
         e.printStackTrace();
      }
   }

   // Helper class to represent the Chat object
   static class Chat {

      @JsonProperty("model")
      private String model;

      @JsonProperty("messages")
      private List<Message> messages;

      @JsonProperty("temperature")
      private float temperature;

      @JsonProperty("max_tokens")
      private int max_tokens;

      @JsonProperty("top_p")
      private float top_p;

      @JsonProperty("frequency_penalty")
      private int frequency_penalty;

      @JsonProperty("presence_penalty")
      private int presence_penalty;
```

```
    public Chat(String model, List<Message> messages,
    float temperature, int max_tokens, float top_p, int
    frequency_penalty, int presence_penalty) {
        this.model = model;
        this.messages = messages;
        this.temperature = temperature;
        this.max_tokens = max_tokens;
        this.top_p = top_p;
        this.frequency_penalty = frequency_penalty;
        this.presence_penalty = presence_penalty;
    }

    // Getters and setters (optional, but can be useful if
        you need to modify the object later)
    }

    // Helper class to represent the Chat Message
    static class Message {
        @JsonProperty("role")
        private String role;

        @JsonProperty("content")
        private String content;

        public Message(String role, String content) {
            this.role = role;
            this.content = content;
        }
    }
}
```

检查前面的代码，你会发现 HTTP 调用是使用 Java API 完成的，没有使用任何外部库——就像我在提示中要求的那样。但请注意，在 Java 中创建和解析 JSON 对象可能会很麻烦，因此我个人指定使用 Jackson API，这一点反映在导入语句和代码本身。

生成的代码包括两个内部类：Chat 和 Message，这两个类可以很容易地分离成单独的 Java 文件。我可以自己手动完成，或者在 ChatGPT 中添加一条新的“用户”信息，要求将内部类分成不同的 Java 文件。

执行 ChatGPTClient.java 后，代码清单 2-9 列出了响应。

代码清单 2-9 响应。调用 ChatGPTClient.java 的结果

```
{
  "id": "chatcmpl-7xIRvjByGobmWH9Vo7OObHCSSwzgI",
  "object": "chat.completion",
  "created": 1694366627,
  "model": "gpt-3.5-turbo-0613",
  "choices": [
    {
      "index": 0,
      "message": {
        "role": "assistant",
        "content": "Java is widely used in the enterprise
primarily due to its numerous benefits and features that make
it a popular choice among large organizations. Here are some
key reasons why Java is so widely adopted in the enterprise:\n\
n1. Platform Independence: One of the biggest advantages of
Java is its platform independence. Java programs can run on
any operating system, making it highly adaptable across a
wide range of devices and platforms. This makes it easier
for enterprises to develop applications that can be deployed
on different systems without any major modifications.\n\n2.
Robustness and Stability: Java is known for its strong emphasis
on reliability, stability, and error handling. It has a built-in
memory management system that prevents memory leaks and
ensures robust performance. This stability is highly valued in
enterprise environments where systems need to run consistently
without disruptions.\n\n3. Scalability: Java offers excellent
scalability, making it suitable for large-scale enterprise
applications. It provides robust support for multi-threading,
allowing applications to handle a large number of concurrent
users smoothly. Java's ability to handle high traffic loads and
distribute processing across multiple servers makes it ideal
for enterprise-level systems.\n\n4. Rich Standard Library and
Frameworks: Java comes with a comprehensive standard library,
offering a wide range of pre-built functions and classes that
simplify development. Additionally, Java has a"
    },
    "finish_reason": "length"
  }
```

```
  ],
  "usage": {
    "prompt_tokens": 28,
    "completion_tokens": 256,
    "total_tokens": 284
  }
}
```

在上面的代码中，你是否注意到对我的提示的回答被截断了？“Additionally，Java has a”并非一个完整句子，原因在于我要求在响应中使用的词元不超过 256 个。它没有超出这个限制。

2.6 小结

与人们普遍认为的相反，ChatGPT 并不掌握读心术！并不具备取代开发人员和架构师的能力，因为它是人工智能。只需要提出一个问题，就能立即得到直接的响应。ChatGPT 无疑可用来将自然语言提示(或请求)转换成代码，但你绝对需要一个开发人员来判断是否应该使用、改进或完全忽略由此产生的代码。

第3章

在企业中使用人工智能！为 Slack 消息创建文本摘要器

在当今的企业世界中，拥有一个 Slack(或 Microsoft 团队)实例，并将其作为与公司每个人沟通的中心场所是非常普遍的。如果你以前用过 Slack，你就会知道，只要公司或世界上某个地方发生了某件重要的事情，一个频道很容易被大量信息淹没。

当然，你在公司承担的责任越大(如经理、团队领导、架构师等)，你需要参与的频道也就越多。在我看来，Slack 是一把双刃剑。你需要用它来完成工作，但作为开发人员，你绝对不能在参加每日例会时说："昨天，呃，我一整天都在看 Slack"。

此外，如果你在一家公司工作，而客户分布在不同时区(这在当今很常见)，早上打开 Slack，看到一大堆在你不在电脑旁边时所收到的消息，这将是非常令人生畏的。

因此，在本章中，我们将在企业中应用人工智能，让 Slack 变得更有用。我们将利用第 2 章中的代码，用 Java 创建一个 Slack 机器人，它将总结 Slack 频道中的重要对话。我们将利用 ChatGPT 的文本摘要功能，并将重点放在"提示工程"上。

3.1 什么是提示工程

简单地说，提示工程就是精心制作、完善提示以及输入参数，以指导和引导 ChatGPT 和其他 AI 模型行为的过程。它基本上是一个行业术语，指的是创建正确的输入以获得你想要的结果。

不过，在继续之前，让我们改进一下第 2 章中的 ChatGPTClient.java。

3.2 使用构建器模式更新 ChatGPTClient.java 及相关类

在第 2 章中，我们创建了 ChatGPTClient.java 作为基本应用程序，用于向聊天端点发送提示信息。这是一个不错的开始，但肯定有改进的余地。

让我们先来看看 Chat 对象的构造函数，它为发送到聊天端点的 JSON Chat 对象建模，如代码清单 3-1 所示。

代码清单 3-1 Chat 对象的构造函数

```
public Chat(String model, List<Message> messages, float
temperature,
            int max_tokens, float top_p, int frequency_
            penalty,
            int presence_penalty) {
            this.model = model;
            this.messages = messages;
            this.temperature = temperature;
            this.max_tokens = max_tokens;
            this.top_p = top_p;
            this.frequency_penalty = frequency_penalty;
            this.presence_penalty = presence_penalty;
        }
```

因此，如果你回顾一下第 2 章中的表 2-4，就会发现，要想成功调用聊天端点，只有 model(模型)和 messages(消息)参数是必需的，而其他参数都是可选的。如果不指定任何参数，其中一些参数会有自己的内置默认值。

这就是我们不需要为整个 Chat JSON 对象“建模”的原因。

所以，这个构造函数基本上需要使用构建器模式进行重构。构建器模式使我们能够获得想要的对象实例，同时只指定我们关心的参数。

此外，Chat(聊天)和 Message(消息)对象不再是内部类，而是存在于它们自己的.java 文件中，这也是合情合理的。代码清单 3-2 展示了如何获取用构建器模式修改过的 Chat 对象实例。

代码清单 3-2　获取 Chat 对象实例

```
Chat chat = Chat.builder()
    .model(model)
    .messages(messages)
    .temperature(temperature)
    .maxTokens(max_tokens)
    .topP(top_p)
    .frequencyPenalty(frequency_penalty)
    .presencePenalty(presence_penalty)
    .build();
```

由于向聊天端点发出的请求必须在聊天 JSON 对象中指定模型和消息，因此我们在 Chat.java 类中添加了一些默认值，使其使用起来更安全(安全指的是该类的用户不易出错)。代码清单 3-3 是新的 Chat.java 文件。

代码清单 3-3　使用构建器模式的 Chat.java

```
import java.util.List;
import java.util.ArrayList;
import com.fasterxml.jackson.annotation.JsonProperty;

public class Chat {
    @JsonProperty("model")
    private String model;

    @JsonProperty("messages")
    private List<Message> messages;

    @JsonProperty("temperature")
    private float temperature;

    @JsonProperty("max_tokens")
    private int max_tokens;
```

```
@JsonProperty("top_p")
private float top_p;

@JsonProperty("frequency_penalty")
private int frequency_penalty;

@JsonProperty("presence_penalty")
private int presence_penalty;
private Chat(ChatBuilder builder) {
   this.model = builder.model;
   this.messages = builder.messages;
   this.temperature = builder.temperature;
   this.max_tokens = builder.max_tokens;
   this.top_p = builder.top_p;
   this.frequency_penalty = builder.frequency_penalty;
   this.presence_penalty = builder.presence_penalty;
}

public static ChatBuilder builder() {

   // we need a default message here to avoid 400 errors
      from the API
   List<Message> messages = new ArrayList<>();
   messages.add(new Message("system", "You are a helpful
   assistant"));
   messages.add(new Message("user", "hello"));

   return new ChatBuilder().messages(messages);
}

public static class ChatBuilder {
   private String model = "gpt-3.5-turbo";
   private List<Message> messages = null;
   private float temperature = 1.0f;
   private int max_tokens = 2048;
   private float top_p = 0f;
   private int frequency_penalty = 0;
   private int presence_penalty = 0;

   private ChatBuilder() {
   }

   public ChatBuilder model(String model) {
      this.model = model;
      return this;
```

```
        }

        public ChatBuilder messages(List<Message> messages) {
            this.messages = messages;
            return this;
        }

        public ChatBuilder temperature(float temperature) {
            this.temperature = temperature;
            return this;
        }

        public ChatBuilder maxTokens(int max_tokens) {
            this.max_tokens = max_tokens;
            return this;
        }

        public ChatBuilder topP(float top_p) {
            this.top_p = top_p;
            return this;
        }

        public ChatBuilder frequencyPenalty(int frequency_
        penalty) {
            this.frequency_penalty = frequency_penalty;
            return this;
        }

        public ChatBuilder presencePenalty(int presence_
        penalty) {
            this.presence_penalty = presence_penalty;
            return this;
        }

        public Chat build() {
            return new Chat(this);
        }
    }
}
```

该类(采用这种设计模式)非常灵活，你可以添加或删除调用聊天端点所需的参数。如果 OpenAI 随时为聊天端点添加新的参数和功能，你可以修改该类以支持新的要求。

为完整起见，代码清单 3-4 显示了 Message.java 类。

代码清单 3-4 Message.java

```
import com.fasterxml.jackson.annotation.JsonProperty;

public class Message {
    @JsonProperty("role")
    private String role;

    @JsonProperty("content")
    private String content;

    public Message(String role, String content) {
        this.role = role;
        this.content = content;
    }
}
```

3.3 ChatGPT 将夺走所有人的工作吗?

我的拙见是，世界上每家公司都坐拥一座尚未开发的信息金矿。如果你拥有记录了员工之间的交流日志的系统、客户支持请求数据库或任何大型文本仓库(是的，这包括你的电子邮件、微软 Exchange 和公司 Gmail)，你就拥有一个待利用的大型非结构化文本仓库。

因此，ChatGPT 的最佳用途不是剥夺任何人的工作，而应该用来增强和扩展公司团队成员已经在做的工作。正如我们在第 2 章中所看到的，作为一名程序员，ChatGPT 可以发挥非常有效的“结对程序员”(Pair-Programmer)作用。它还能高效、快速地完成某些困难的任务。下面，我们来看一个实际例子，看看如何让大量非结构化文本变得有用。

3.4 研究一个真实世界的问题:软件公司的客户支持

软件开发中最艰巨的任务之一就是提供技术支持。试想，如果有人在使用你的软件时感到沮丧、困惑或有需要解决的问题，那么你必须及时进行处理。以下是客户支持工作难做的一些原因：

- 你的最终用户和客户不擅于解释软件问题。
- 初级技术人员通常是第一道防线，一般处理最基本的问题或用户错误。但当问题变得更加复杂时，就会升级到中级。
- 中级是一个棘手的地方。他们比一级技术支持人员拥有更多的知识和经验；但是，他们没有机会直接从最终用户那里获得答案。
- 真正糟糕的问题会升到高级；然而，这些技术支持人员的成本最高，因为他们拥有最多的知识和经验，比如代码、服务器和基础设施方面的实践经验。

因此，让我们以 Slack 中一个典型的技术支持频道中的典型对话为例进行分析。下面列出一个虚构公司的团队成员及其角色：

- Fatima(客户服务代表)
- John(软件工程师)
- Dave(项目经理)
- Keith(首席技术官)

代码清单 3-5 提供了一个软件初创公司团队成员之间对话的示例。客户服务代表 Fatima 告诉团队，他们的应用程序在启动后立即崩溃(这可不是一个小问题)。首席技术官 Keith 立即介入，将问题升级。

代码清单 3-5　试图分析客户问题的 Slack 频道内的团队成员

```
Fatima [16:00 | 02/08/2019]: Hey everyone, I have an urgent
issue to discuss. I just got off a call with a client who's
experiencing app crashes as soon as they load it. They're
really frustrated. Can we get this sorted ASAP? 😫

Keith [16:01 | 02/08/2019]: Thanks for bringing this to our
attention, Fatima. Let's jump on this right away. @John,
can you take the lead in investigating the issue since our
architect is out sick today?

John [16:02 | 02/08/2019]: Sure thing, Keith. I'll dive into
the codebase and see if I can find any potential culprits for
the crashes.

John [16:02 | 02/08/2019]: Fatima, could you gather some
additional information from the client? Ask them about the
specific device, operating system, and any recent updates they
```

```
might have installed.

Fatima [16:03 | 02/08/2019]: Absolutely, John. I'll reach out
to the client immediately and gather those details. Will update
you all once I have them.

Dave [16:04 | 02/08/2019]: I understand the urgency here. Let's
make sure we keep the client informed about our progress 😩
Fatima. We don't want them feeling left in the dark during this
troubleshooting process.

Fatima [16:04 | 02/08/2019]: Definitely, Dave. 👍 I'll keep the
client updated at regular intervals, providing them with any
relevant information we uncover.

John [16:20 | 02/08/2019]: I've checked the codebase, and so
far, I haven't found any obvious issues. It's strange that the
app is crashing on load. Could it be a memory-related issue?
Keith, do we have any recent reports of memory leaks or high
memory usage?

Keith [16:22 | 02/08/2019]: I'll pull up the monitoring logs,
John, and check if there have been any memory-related anomalies
in recent releases. Let me get back to you on that.

Fatima [17:01 | 02/08/2019]: Quick update, everyone. The client
is using an iPhone X running iOS 15.1. They mentioned that the
issue started after updating their app a few days ago 😲

Keith [17:05 | 02/08/2019]: Thanks for the update, Fatima.
That's helpful information. John, let's focus on testing the
latest app update on an iPhone X simulator with iOS 15.1 to see
if we can replicate the issue.

John [17:06 | 02/08/2019]: Good idea, Keith. I'll set up the
emulator and run some tests right away.

Keith [17:30 | 02/08/2019]: John, any progress on replicating
the issue on the emulator?

John [17:32 | 02/08/2019]: Yes, Keith. I managed to reproduce
the crash on the emulator. It seems to be related to a
compatibility issue with iOS 15.1 🤓. I suspect it's due to a
deprecated method call. I'll fix it and run more tests to
confirm.
```

```
John [18:03 | 02/08/2019]: Fixed the deprecated method issue,
and the app is no longer crashing on load. It looks like we've
identified and resolved the problem. I'll prepare a patch and
send it to you, Keith, for review and deployment.

Keith [18:04 | 02/08/2019]: 👏👏👏 Thank you, please provide
me with the patch as soon as possible. Once I review it, we'll
deploy the fix to the app store.

Dave [18:06 | 02/08/2019]: Great job, team! 🎉 John, please
keep the client informed about the progress and let them know
we have a fix ready for them on the next app update. Can
someone make sure the release notes reflect this?

John [18:07 | 02/08/2019]: Will do, Dave. I'll update the
client and ensure they're aware of the upcoming fix.

Keith [18:27 | 02/08/2019]: Patch reviewed and approved, John.
Please proceed with updating the app in the store. Let's aim to
have it done within the next hour.

John [18:26 | 02/08/2019]: Understood, Keith. I'm in the
process of uploading it now.

Fatima [18:38 | 02/08/2019]: I just informed the client about
the fix. They're relieved and grateful for our prompt response.
Thanks, everyone, for your collaboration and quick action. It's
a pleasure working with such a competent team!

Dave [18:40 | 02/08/2019]: Well done, team! Your efforts are
greatly appreciated. We managed to turn this urgent problem
around in record time. Let's keep up the good work!
```

3.5 提示工程入门：文本摘要

不用说，没有人愿意整天不停地浏览 Slack 频道，阅读那些轰动性的议题和问题。我们将利用 ChatGPT 的功能获得文本摘要。为了简单起见，我们先试着用几条提示将聊天信息的整个列表发送给 ChatGPT，以便它为我们提供所有已发生事件的可用摘要。

3.5.1 提示 1：tl;dr

tl;dr 这条提示简短、动听、直奔主题。它是网上常用的一个术语，用来概括一篇长文章。它的字面意思是“太长；未读(Too Long; Didn't Read)”，这也是当我忙于工作时对 Slack 上很多对话的感受。ChatGPT 可以理解这个简单的术语(我不一定称它为单词)作为提示，而且效果相当不错。

代码清单 3-6 提示。要求较长对话的摘要版

```
System: You are a helpful assistant
User: Fatima [16:00 | 02/08/2019]: Hey everyone, I have an
urgent ...
User: tl;dr
Model: gpt-4
Temperature: 1
Maximum length: 360
Top P: 1
Frequency penalty: 0
Presence penalty: 0
```

中文注释

系统：你是一个乐于助人的助手

用户：Fatima[16:00|02/08/2019]：大家好，我遇到紧急情况……

用户：tl;dr

模型：GPT-4

温度：1

最大长度：360

Top P：1

频率惩罚：0

存在惩罚：0

注意：

关于如何向 ChatGPT 发送前面的提示，以下两个选择供参考。第一是使用 ChatGPTClient.java 应用程序及其支持类，第二是使用 Chat Playground。无论哪种情况，都要确保第一条用户信息包含代码清单 3-5 中的所有文本。代码清单 3-6 有删节，使其更易读。

请注意词元数保持为 360 这个较小值，这将最终帮助 ChatGPT 创建一个简短响应。还可看到温度保持在 1(默认值)，Top P 却一直增加到最大值(即 1)。因此，使用温度和其他设置来控制 ChatGPT 响应的语气和多样性十分重要。

代码清单 3-7 是 ChatGPT 对代码清单 3-5 中对话的摘要。

代码清单 3-7　响应。获取冗长对话的摘要版

```
The team quickly responds to a client's app crashing issue.
John investigates the codebase while Fatima gathers information
from the client. They discover a compatibility issue with iOS
15.1 and John fixes it. The patch is reviewed and approved,
and the app is updated in the store. The client is informed
and grateful for the quick resolution. The team is praised for
their collaboration and prompt action.
```

团队快速响应客户的应用程序崩溃问题。John 调查代码库，Fatima 从客户处收集信息。他们发现了 iOS 15.1 的兼容性问题，John 对其进行了修复，补丁得到了审核和批准，应用程序也在商店中进行了更新。客户得到了通知，并对快速解决问题表示感谢。团队的协作和迅速行动得到了赞扬。

这真是太棒了！它将极其冗长的对话缩减为只有 68 个字的简要总结。正如你所看到的，ChatGPT 在这里并不能完全取代人的工作；但是，它肯定会让人的工作变得更有效率。现在，让我们再尝试几个提示！

3.5.2 提示 2：“用 3 句或更少的话解释这个问题”

如果“tl;dr”不能为你提供足够简短的解释，那么我们可以告诉 ChatGPT 让事情变得更简单。ChatGPT 在参数方面做得很好，所以我们指定解释不能超过 3 句话，如代码清单 3-8 所示。

代码清单 3-8 提示。要求较长对话的简短版

```
System: You are a helpful assistant
User: Fatima [16:00 | 02/08/2019]: Hey everyone, I have an
urgent ...
User: Explain this in 3 sentences or less.
Model: gpt-4
Temperature: 1
Maximum length: 250
Top P: 0
Frequency penalty: 0
Presence penalty: 0
```

中文注释

系统：你是一位乐于助人的助手

用户：Fatima[16:00|02/08/2019]：大家好，我遇到紧急情况……

用户：请解释一下，不超过 3 句话。

模型：GPT-4

温度：1

最大长度：250

Top P：0

频率惩罚：0

存在惩罚：0

我们非常希望 ChatGPT 能给出一个简洁答案，所以温度越“干”越好。

代码清单 3-9　响应。获取不超过 3 句话的摘要

```
A customer reported an urgent issue of app crashes to the
team, John investigates the issue and asks Fatima to gather
additional information from the client. They discover that the
issue is related to a compatibility problem with iOS 15.1. John
fixes the problem and Keith approves the deployment of the fix
to the client's production environment.
```

中文注释

一位客户向团队报告了一个应用程序崩溃的紧急问题，John 调查了这个问题，并要求 Fatima 从客户那里收集更多信息。他们发现该问题与 iOS 15.1 的兼容性问题有关。John 修复了问题，Keith 批准将修复程序部署到客户的生产环境中。

3.5.3　提示 3："我是一名经理。向我解释一下发生了什么情况"？

ChatGPT 可以重新修改措辞，并将其拆分为非常简单的内容(代码清单 3-10)。

代码清单 3-10　提示。要求提供更复杂对话的简化版

```
System: You are a helpful assistant
User: Fatima [16:00 | 02/08/2019]: Hey everyone, I have an
urgent ...
User: Summarize this conversation and explain it to me like I'm
a manager with little technical experience.
Model: gpt-3.5-turbo
Temperature: 0.82
Maximum length: 750
Top P: 1
Frequency penalty: 0
Presence penalty: 0
```

中文注释

系统：你是一位乐于助人的助手

用户：Fatima[16:00|02/08/2019]：大家好，我遇到紧急情况……

用户：请总结一下这段对话，并向我解释一下，当我是一个没有什么技术经验的经理。

模型：GPT-3.5-TURBO

温度：0.82

最大长度：750

Top P：1

频率惩罚：0

存在惩罚：0

我们将温度设置在 1.0 以下，以便 ChatGPT 的答案能够保持简明易懂。我们增加了词元长度，以便 ChatGPT 能花时间解释一些潜在的技术概念。

代码清单 3-11　响应。获取更复杂对话的简化版

```
In this conversation, Fatima brings up an urgent issue with a
client who is experiencing app crashes as soon as they load
it. Keith assigns John to investigate the issue since their
architect is out sick. John asks Fatima to gather additional
information from the client about their device, operating
system, and recent updates. Dave emphasizes the importance of
keeping the client informed throughout the troubleshooting
process.

John checks the codebase but doesn't find any obvious issues,
so he suspects it could be a memory-related problem. Keith
agrees to check for memory-related anomalies in recent
releases. Fatima provides an update that the client is using an
iPhone X running iOS 15.1 and mentioned that the issue started
after updating the app. Keith suggests testing the latest app
update on an iPhone X simulator with iOS 15.1 to replicate
```

```
the issue.

John successfully reproduces the crash on the emulator and
identifies a compatibility issue with iOS 15.1, possibly due
to a deprecated method call. He fixes the issue and prepares a
patch for review and deployment. Keith approves the patch and
asks John to update the app in the store. Dave requests that
the client be kept informed and that the release notes reflect
the upcoming fix.

John uploads the patch and informs the client about the fix.
Keith reviews and approves the patch, and John proceeds with
updating the app in the store. Fatima informs the client about
the fix, and they express relief and gratitude for the prompt
response. Dave commends the team for their collaboration
and quick action, emphasizing the importance of maintaining
good work.

In summary, the team quickly addressed an urgent issue where
a client's app was crashing on load. They investigated the
issue, identified a compatibility problem with iOS 15.1, fixed
it, and prepared a patch for deployment. The client was kept
informed throughout the process, and the fix was successfully
implemented, resulting in a relieved and grateful client.
```

中文注释

在这段对话中，Fatima 向客户提出了一个紧急问题，客户的应用程序一加载就会崩溃。Keith 指派 John 去调查这个问题，因为他们的架构师请了病假。John 要求 Fatima 从客户那里收集更多关于设备、操作系统和最近更新的信息。Dave 强调了在整个故障排除过程中让客户知情的重要性。

John 仔细检查了代码库，但并未发现明显的问题，因此他推测问题可能与内存有关。Keith 也认同这一观点，并同意检查最近版本中与内存相关的异常。此时，Fatima 提供了新的线索，客户使用的是运行 iOS 15.1 的 iPhone X，并且问题是在更新应用程序后才出现的。于是，Keith 提出在模拟 iOS 15.1 的 iPhone X 环境中测试最新的应用程序更新，以尝试复现这个问题。

经过努力，John 成功地在模拟器上重现了崩溃现象，并确定问题与 iOS 15.1 的兼容性问题有关，很可能是由于调用了过时的方法。他迅速修复了这个问题，并准备了一个补丁供团队审核和部署。Keith 审查并批准了补丁，并要求 John 尽快更新商店中的应用程序。Dave 也强调，务必及时通知客户，并在发布说明中明确注明即将进行的修复。

随后，John 上传了补丁，并向客户发送了修复通知。Keith 再次审查并批准了补丁，John 便着手更新商店中的应用程序。同时，Fatima 也向客户通报了修复情况，客户对公司的迅速响应表示非常欣慰和感谢。Dave 对团队的协作和快速行动表示赞赏，并强调了保持高效沟通和协作的重要性。

总之，团队迅速而有效地解决了客户应用程序在加载时崩溃的紧急问题。他们深入调查，确定了与 iOS 15.1 的兼容性问题，迅速修复并准备了补丁。在整个过程中，他们始终与客户保持沟通，并成功实施了修复，让客户备感安心和感激。

3.5.4 提示 4："给我下一步建议"

除了做出更容易理解的摘要，ChatGPT 还能以不同的风格和"语气"进行解释。我们将把 Top P 值改为 0.75，因为我们希望 ChatGPT 能想到比默认值更多的可能性。我们还要将存在惩罚提高到 0.5，以鼓励多方面的回答。

代码清单 3-12 提示。向 ChatGPT 寻求最佳下一步建议

```
System: You are a helpful assistant
User: Fatima [16:00 | 02/08/2019]: Hey everyone, I have an
urgent ...
User: I'm a QA engineer at the company. Give me suggestions on
next steps.
Model: gpt-4
Temperature: 1
Maximum length: 1200
Top P: 0.75
Frequency penalty: 0
Presence penalty: 0.5
```

中文注释

系统：你是一位乐于助人的助手

用户：Fatima[16:00|02/08/2019]：大家好，我遇到紧急情况……

用户：我是公司的 QA 工程师。给我下一步的建议。
模型：GPT-4
温度：1
最大长度：1200
Top P：0.75
频率惩罚：0
存在惩罚：0.5

代码清单 3-13　响应。ChatGPT 就下一步可能采取的措施提出的建议

```
As a QA engineer, here are some suggestions for next steps:

1. Test the patch: Once the patch has been deployed to the app
store, thoroughly test it on various devices and operating
systems to ensure that the issue has been resolved and there
are no new bugs or regressions introduced.

2. Conduct regression testing: In addition to testing the
specific issue, perform a comprehensive regression test to
ensure that the fix does not have any unintended side effects
on other functionalities of the app.

3. Communicate the fix to the client: Keep the client updated
on the progress and let them know that the fix is available in
the latest app update. Provide them with clear instructions on
how to update their app and encourage them to test it on their
end as well.

4. Monitor user feedback: Keep an eye on user feedback after
the fix has been deployed. Encourage users to report any issues
they encounter, and be proactive in addressing them promptly.

5. Document the issue and resolution: Create a detailed report
documenting the issue, the steps taken to reproduce and resolve
it, and any other relevant information. This will serve as
a reference for future troubleshooting and help improve the
development process.

6. Conduct a post-mortem analysis: After the issue has been
resolved and the fix has been deployed, schedule a post-mortem
```

```
meeting with the team to discuss the root cause, lessons
learned, and any process improvements that can be implemented
to prevent similar issues in the future.

7. Continuous improvement: Use this incident as an opportunity
to identify areas of improvement in the development and testing
processes. Evaluate if there are any changes that can be made
to prevent similar issues from occurring in the future.

Remember, open communication, collaboration, and continuous
learning are key to ensuring the highest quality of your app
and maintaining a strong relationship with your clients.
```

中文注释

作为一名 QA 工程师，下面有一些关于下一步工作的建议。

1. 测试补丁：将补丁部署到应用程序商店后，在各种设备和操作系统上进行彻底测试，以确保问题已得到解决，并且没有引入新的错误或回归。

2. 进行回归测试：除了测试具体问题外，还要进行全面的回归测试，以确保修复程序不会对应用程序的其他功能产生任何意想不到的副作用。

3. 向客户通报修复情况：随时向客户通报最新进展，让他们知道修复程序已在最新的应用程序更新中提供。向他们提供如何更新应用程序的明确说明，并鼓励他们在自己的终端进行测试。

4. 监控用户反馈：部署修复程序后，密切关注用户反馈。鼓励用户报告他们遇到的任何问题，并积极主动地及时解决这些问题。

5. 记录问题和解决方案：创建一份详细的报告，记录问题、重现和解决问题的步骤以及任何其他相关信息。这将作为今后排除故障的参考，并有助于改进开发流程。

6. 进行事后分析：在问题得到解决并部署了修复程序后，与团队安排一次事后分析会议，讨论问题的根本起因、吸取的经验教训以及可实施的任何流程改进，以防今后出现类似问题。

7. 持续改进：以此次事件为契机，确定开发和测试流程中需要改进的地方。评估是否可以做出任何改变，以防止今后发生类似问题。

请记住，开放式沟通、协作和持续学习是确保应用程序质量上乘、维系良好客户关系的关键所在。

当然，ChatGPT(如同全球其他形式的人工智能一样)并非尽善尽美。例如，尽管第 3 项建议看似有效，但实际上，直接与客户或顾客沟通通常并非 QA 人员的核心职责。这种沟通更适合由技术支持团队或拥有相应沟通渠道的产品经理来承担，特别是当涉及重要客户时。因此，尽管这个建议有其可取之处，但对于在公司中担任 QA 角色的人来说，可能并不完全适用。

3.5.5　深入探讨提示工程

当你在谷歌上搜索“提示工程”这个词时，会发现大量的示例、博客，甚至是提供订阅计划的完整网站，它们都在试图通过文本输入来打造完美的提示。然而，正如你从上例中看到的，提示工程远非仅仅制作文本输入那么简单。

实际上，这个过程与烹饪一道精致的菜肴颇为相似。想象一下，如果只用盐作为调味料(完全不考虑其他配料和香料)来烹饪红酒牛肉，结果必然是单调乏味的，与真正的佳肴相去甚远。

同样，试图用单一乐器和一名乐手组建一个完整的交响乐团，这样的“单人乐队”无疑是令人尴尬的。因此，要真正完成提示工程，仅仅根据提示调整文本是远远不够的。模型的温度(控制随机性)、Top P(影响词元概率)、使用的特定模型、词元数量以及端点的其他参数，都对获得出色的响应起着至关重要的作用。

虽然本书并非专门探讨提示工程，但从上面的解释中我们可以看出，它确实涉及多个与 Java 无关的因素。不过，我们强烈建议你充分利用 OpenAI 提供的模型和端点的所有参数，以找到最适合使用场景的方法。

3.6　注册 Slack Bot 应用程序

既然我们已经知道了 ChatGPT 总结大量文本的各种方法，那么接下来看看用 Java 创建一个简单的机器人需要做些什么，这个机器人将以编程方式从 Slack 实例频道抓取所有消息。

注意：

要完成这些步骤，你需要拥有 Slack 工作区的管理权限。大多数开发人员都没有这些级别的权限；因此，为了充分进行试验，建议你创建自己的个人 Slack 工作区进行测试。这样，你就拥有了安装 Slack 机器人的所有权限。

但要一步一步来。首先，我们要制作 Slack 机器人应用程序，请访问 Slack API 网站(图 3-1)，网址是 https://api.slack.com/。

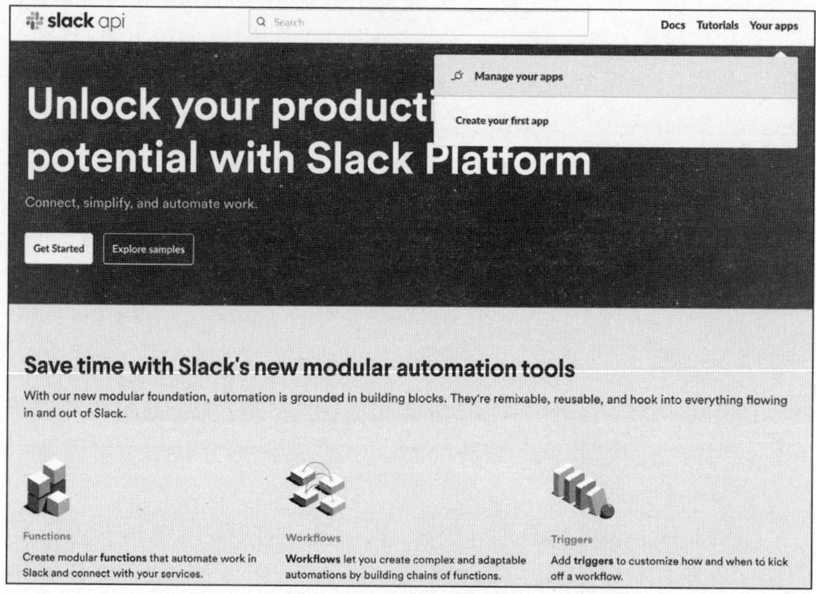

图 3-1 访问 Slack API 网站创建 Slack 机器人

当然，你需要有一个 Slack 账户才能使用此功能；如果没有账户，就需要先创建一个。

登录后，转到页面右上方，导航到 Your apps | Create your first app，如图 3-1 所示。在 Slack 术语中，“机器人”就是“应用程序”，除非先在 Slack 注册，否则无法在 Slack 实例上运行机器人。

如图 3-2 所示，你将进入 Your Apps 页面，在这里可以管理你的 Slack 应用程序。随即，你会在屏幕中间看到弹出的 Create an app 按钮。

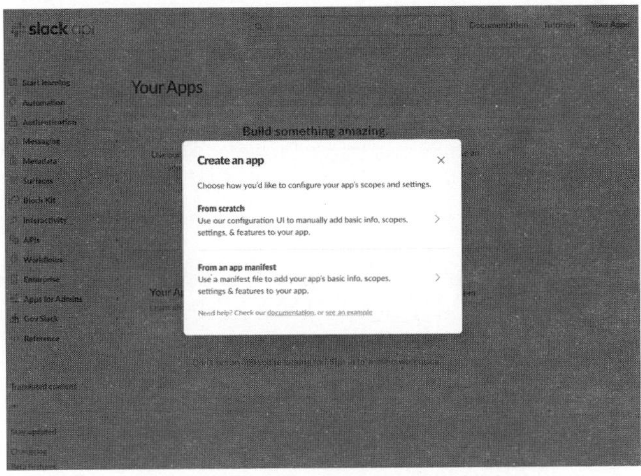

图 3-2　为 Slack 创建新的机器人应用程序

选择 From scratch 选项，因为我们希望能够自己操作应用程序的所有细节，而不希望使用一堆默认设置，让事情变得过于复杂。

随后，系统会提示你为机器人指定一个名称，并选择你希望机器人访问的工作区，如图 3-3 所示。

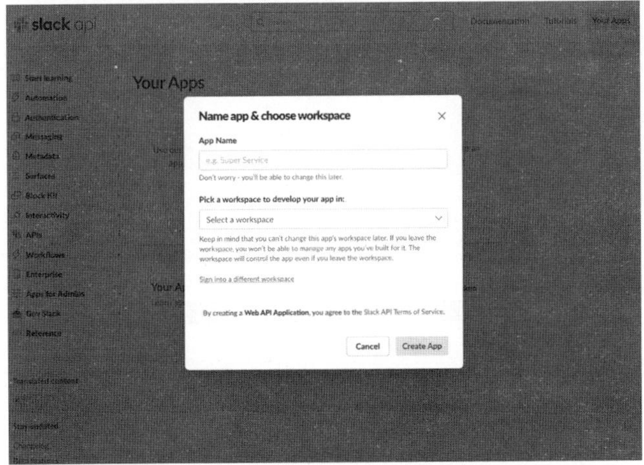

图 3-3　为 Slack 创建新的机器人应用程序

单击 Create App 按钮继续下一步。

3.6.1 通过设置范围指定机器人的权限

现在，你将看到一个屏幕，上面有大量 Slack 工作区机器人的选项。不过，你需要做的第一件事是从左侧边栏单击 OAuth & Permissions。

机器人将非常简单；它需要做的就是读取频道中的消息，以便为我们提供内容摘要。除了读取消息，还需要知道 Slack 工作区中人员的姓名；否则，得到的将是人员的 UUID 表示，而非姓名，这对我们来说毫无意义。

因此，请向下滚动并确保将以下 OAuth 作用域添加到 Slack Bot 中，如图 3-4 所示。

- channels: history(频道：历史)
- channels: read(频道：读取)
- users: read(用户：读取)

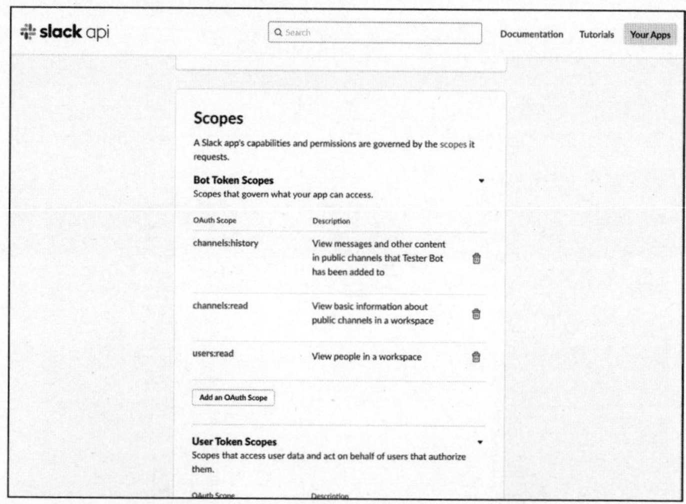

图 3-4 为 Slack Bot 应用程序添加作用域

3.6.2 确认设置

为机器人添加适当的作用域后，向上滚动并单击左侧栏中的 Basic Information。

在接下来的页面中，你会看到 Add features and functionality 旁边有一

个绿色的复选标记，这表明你已经正确添加了作用域，如图 3-5 所示。

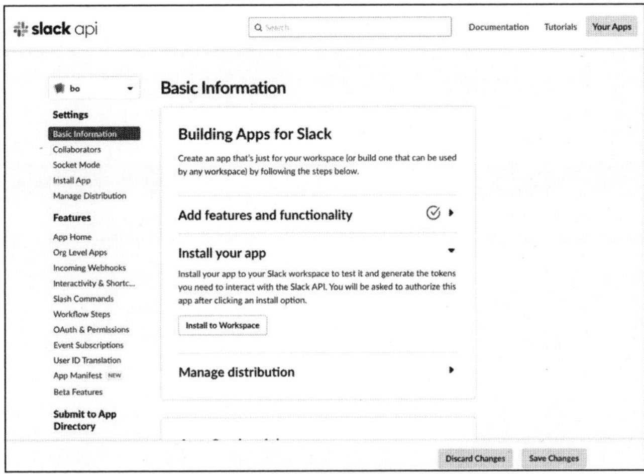

图 3-5 确认设置

3.6.3 查看 OAuth & Permissions 页面

如图 3-6 所示，导航到 OAuth & Permissions 页面，然后单击 Install to Workspace 按钮。

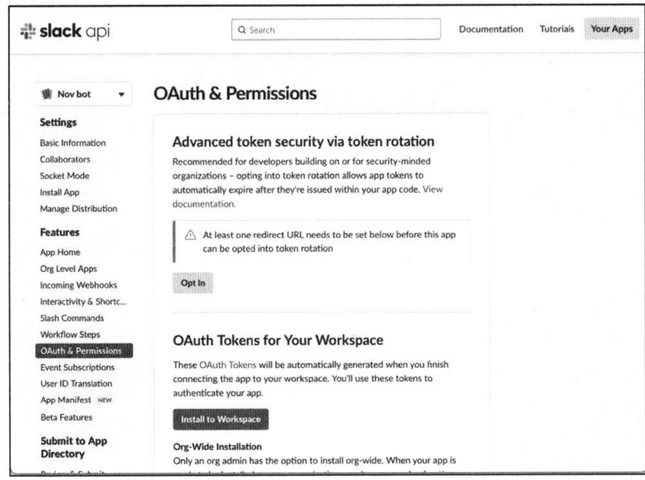

图 3-6 OAuth & Permissions 页面

3.6.4 将 Slack Bot 应用程序安装到工作区

现在，所有权限都已申请完毕，是时候将机器人安装到工作区了。在安装过程中，你应该会看到如图 3-7 所示的界面。

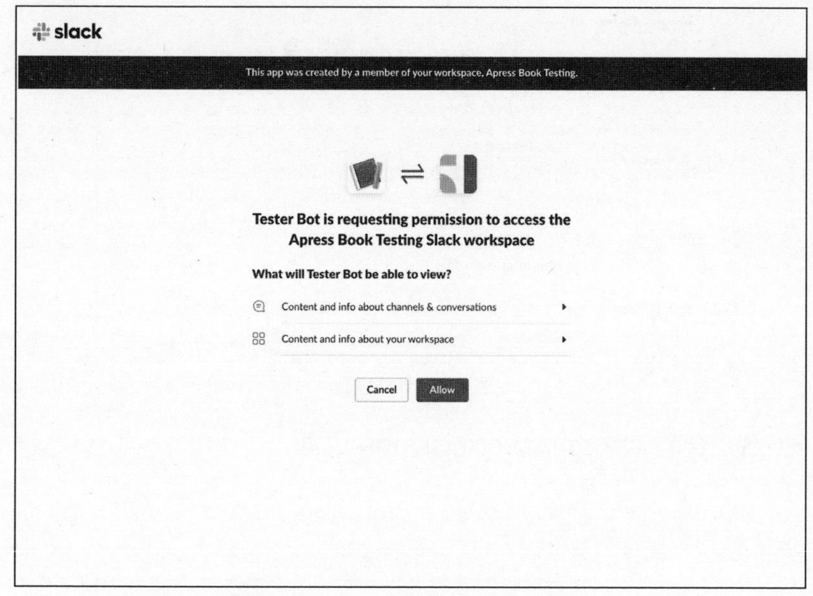

图 3-7 “安装”新的 Slack Bot 程序

单击 Allow 按钮，对机器人进行授权，并获得你在上一步中分配的权限。

注意：

理解这里的“安装”含义很重要。在传统的 Java 意义上，安装应用程序意味着将 JAR、WAR 或 EAR 文件加载到另一台机器并让其执行。这里的情况并非如此。

这里，当你“安装”一个机器人应用程序时，将启用 Slack 工作区，允许一个应用程序加入工作区，仅此而已。机器人的代码将运行在你自己的机器上，而非 Slack 的服务器上。

3.6.5　获取 Slack 机器人访问词元

这次，“词元”实际上指的是访问词元！为了连接到 Slack API 并以编程方式访问消息和用户信息，你需要为 Slack 机器人生成一个特定的 OAuth 词元。

回到 OAuth & Permissions 页面，确保从这里复制机器人词元(通常以"xoxb-"开头)，如图 3-8 所示。

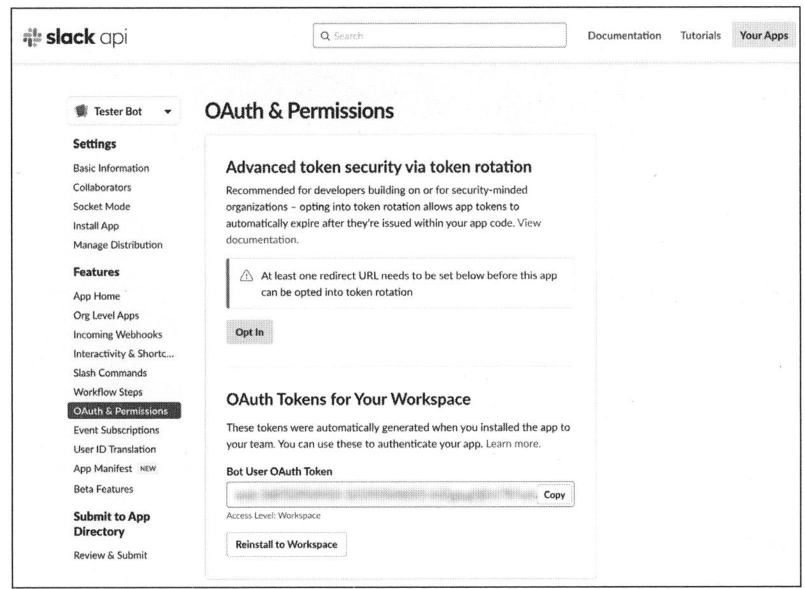

图 3-8　为 Slack 机器人程序复制 OAuth 词元

3.6.6　邀请机器人访问你的频道

接下来，进入你想用来测试机器人的频道，并在频道中键入以下命令。

```
/invite
```

选择 Add apps to this channel 选项，然后选择你之前在 Slack 上注册机器人时指定的 Slack Bot 的名称，如图 3-9 所示。

恭喜你！你现在已经成功地在 Slack 注册了一个 Slack Bot 应用程序，

使其能够读取工作区中的消息，并将 Slack Bot 添加到一个频道。在编写 Java 代码以访问工作区中的频道之前，我们需要知道 Slack 为频道使用的内部 ID 是什么。

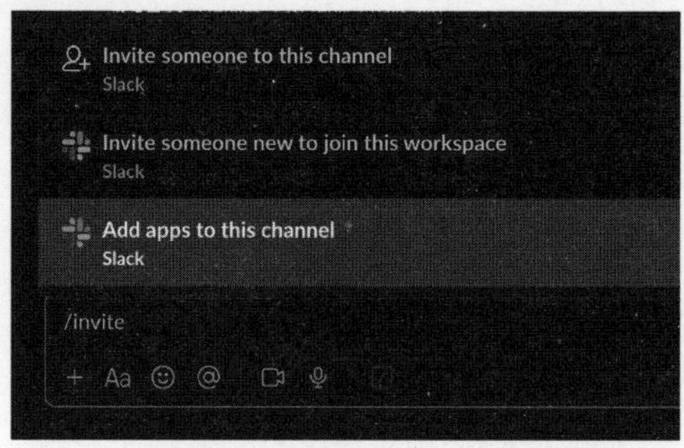

图 3-9 将 Slack Bot 添加到频道

3.7 查找频道 ID

好了，这一步很简单。在 Slack 中，右击频道名称，选择 View channel details 选项。弹出窗口的底部显示了频道的 ID。复制这个数字并保存起来备用，因为你的 Java 应用程序需要这个 ID 才能加入 Slack 工作区中的正确频道。

3.8 使用 Slack Bot 应用程序自动从频道抓取消息

好了，既然已经具备了所有前提条件，也知道了频道的 ID，下面就用 Java 编写代码，访问特定 Slack 频道中的所有消息。

3.8.1 设置依赖关系

Java 的 Slack API 库提供了与 Slack 平台交互的便捷方法。我们需要的大部分内容都来自 com.slack.api.methods.*或 com.slack.api.model.*包，这两个包位于 slack-api-client-<VERSION>和 slack-api-model-<VERSION>jar 文

件中。

Slack Java API 有自己的依赖关系，它们是

- GSON
  - gson-<VERSION>.jar
- Kotlin
  - kotlin-stdlib-<VERSION>.jar
  - kotlin-stdlib-jdk8-<VERSION>.jar
- OK HTTP 和 OK IO
  - okhttp-<VERSION>.jar
  - okio-<VERSION>.jar
  - okio-jvm-<VERSION>.jar
- SL4J
  - slf4j-api-<VERSION>.jar

代码清单 3-14 和代码清单 3-15 是构建所有内容需要的 Maven pom.xml 和 Gradle build.gradle 文件的片段(使用我测试过的版本)。

代码清单 3-14　Maven pom.xml

```
<dependencies>
   <!-- Gson library -->
   <dependency>
      <groupId>com.google.code.gson</groupId>
      <artifactId>gson</artifactId>
      <version>2.10.1</version>
   </dependency>

   <!-- Kotlin standard libraries -->
   <dependency>
      <groupId>org.jetbrains.kotlin</groupId>
      <artifactId>kotlin-stdlib</artifactId>
      <version>1.6.20</version>
   </dependency>
   <dependency>
      <groupId>org.jetbrains.kotlin</groupId>
      <artifactId>kotlin-stdlib-jdk8</artifactId>
      <version>1.6.20</version>
   </dependency>
```

```
    <!-- OkHttp library -->
    <dependency>
        <groupId>com.squareup.okhttp3</groupId>
        <artifactId>okhttp</artifactId>
        <version>4.11.0</version>
    </dependency>

    <!-- Okio library -->
    <dependency>
        <groupId>com.squareup.okio</groupId>
        <artifactId>okio</artifactId>
        <version>3.2.0</version>
    </dependency>
    <dependency>
        <groupId>com.squareup.okio</groupId>
        <artifactId>okio-jvm</artifactId>
        <version>3.2.0</version>
    </dependency>

    <!-- Slack SDK libraries -->
    <dependency>
        <groupId>com.slack.api</groupId>
        <artifactId>slack-api-client</artifactId>
        <version>1.30.0</version>
    </dependency>
    <dependency>
        <groupId>com.slack.api</groupId>
        <artifactId>slack-api-model</artifactId>
        <version>1.30.0</version>
    </dependency>

    <!-- SLF4J logging facade -->
    <dependency>
        <groupId>org.slf4j</groupId>
        <artifactId>slf4j-api</artifactId>
        <version>2.0.7</version>
    </dependency>
</dependencies>
```

代码清单 3-15 Gradle build.gradle

```
dependencies {
    // Gson library
```

```
    implementation 'com.google.code.gson:gson:2.10.1'

    // Kotlin standard libraries
    implementation 'org.jetbrains.kotlin:kotlin-stdlib:1.6.20'
    implementation 'org.jetbrains.kotlin:kotlin-stdlibjdk8:
    1.6.20'

    // OkHttp library
    implementation 'com.squareup.okhttp3:okhttp:4.11.0'

    // Okio library
    implementation 'com.squareup.okio:okio:3.2.0'
    implementation 'com.squareup.okio:okio-jvm:3.2.0'

    // Slack SDK libraries
    implementation 'com.slack.api:slack-api-client:1.30.0'
    implementation 'com.slack.api:slack-api-model:1.30.0'

    // SLF4J logging facade
    implementation 'org.slf4j:slf4j-api:2.0.7'
}

compileKotlin {
    kotlinOptions {
        jvmTarget = "1.8"
    }
}
compileTestKotlin {
    kotlinOptions {
        jvmTarget = "1.8"
    }
}
```

现在我们有了访问词元和所有必需的依赖项，让我们来看看访问频道并抓取指定时间范围内所有聊天记录需要的代码。显而易见，我们需要该频道中每次发布的用户名、时间戳和消息内容。

3.8.2 使用 ChannelReaderSlackBot.java 以编程方式从 Slack 读取消息

代码清单 3-16 是一个简单的 Java Slack Bot，能获取指定时间段内频道中每次发布的用户名、时间戳和消息内容。

代码清单 3-16 ChannelReaderSlackBot.java

```
import com.slack.api.Slack;
import com.slack.api.methods.MethodsClient;
import com.slack.api.methods.request.conversations.
ConversationsHistoryRequest;
import com.slack.api.methods.response.conversations.
ConversationsHistoryResponse;
import com.slack.api.methods.request.users.UsersInfoRequest;
import com.slack.api.methods.response.users.UsersInfoResponse;
import com.slack.api.model.Message;
import com.slack.api.model.User;
import com.slack.api.model.block.LayoutBlock;

import java.time.*;
import java.util.Collections;
import java.util.List;

public class ChannelReaderSlackBot {

        private static final String SLACK_BOT_TOKEN = "YOUR_
        SLACK_API_TOKEN";

        public static void main(String[] args) {
        Slack slack = Slack.getInstance();
        MethodsClient methods = slack.methods(SLACK_BOT_TOKEN);

        String channelId = "YOUR_CHANNEL_ID";

        LocalDateTime startTimeUTC = LocalDateTime.of(2023,
        Month.AUGUST, 3, 10, 0);
        LocalDateTime endTimeUTC = LocalDateTime.of(2023,
        Month.AUGUST, 12, 15, 0);

        long startTime = startTimeUTC.atZone(ZoneOffset.UTC).
        toEpochSecond();

        long endTime = endTimeUTC.atZone(ZoneOffset.UTC).
        toEpochSecond();
        ConversationsHistoryRequest request =
        ConversationsHistoryRequest.builder()
            .channel(channelId)
            .oldest(String.valueOf(startTime))
            .latest(String.valueOf(endTime))
            .build();
```

```
try {
   ConversationsHistoryResponse response = methods.
   conversationsHistory(request);

      if (response != null && response.isOk()) {
         List<Message> messages = response.
         getMessages();
         Collections.reverse(messages);
         for (Message message : messages) {
            String userId = message.getUser();
            String timestamp = formatTimestamp(message.
            getTs());

            UsersInfoRequest userInfoRequest =
            UsersInfoRequest.builder()
               .user(userId)
               .build();

            UsersInfoResponse userInfoResponse =
            methods.usersInfo(userInfoRequest);
            if (userInfoResponse != null &&
            userInfoResponse.isOk()) {
               User user = userInfoResponse.getUser();
               System.out.println("User: " + user.
               getName());
               System.out.println("Timestamp: " +
               timestamp);
               System.out.println("Message: " +
               message.getText());
               System.out.println();
            }
         }
      } else {
         System.out.println("Failed to fetch messages: "
         + response.getError());
      }
   } catch (Exception e) {
      e.printStackTrace();
   }
}

private static String formatTimestamp(String ts) {
   double timestamp = Double.parseDouble(ts);
   Instant instant = Instant.ofEpochSecond((long)
```

```
            timestamp);
            LocalDateTime dateTime = LocalDateTime.
            ofInstant(instant, ZoneOffset.UTC);
            return dateTime.toString();
        }
    }
```

当然，你应该将 YOUR_SLACK_API_TOKEN 替换为实际的 Slack API 词元，并将 YOUR_CHANNEL_ID 替换为你要从其中读取消息的 Slack 频道的 ID。

如果你想完成一些非常基本的事情，使用 Slack Java API 其实非常简单。如果不首先使用静态调用 Slack.getInstance()方法获得 Slack 类的实例，就无法执行任何操作。这将连接到底层的 Slack API 基础架构，让你可以与公开方法交互，获取想要的信息。

接下来，需要通过调用 slack.methods()方法获得 MethodsClient 类的实例，并提供访问词元。

为检索聊天历史记录，我们使用 ConversationsHistoryRequest 类，这是 Slack API 提供的另一个类。在这里，你只需要指定所需的频道 ID、最旧的时间戳和最新的时间戳，即可定义聊天历史记录的时间范围。在本例中，将检索 2023 年 8 月 3 日 10:00 至 2023 年 8 月 12 日 15:00 的消息，简直轻而易举。

代码清单 3-17 显示了执行 ChannelReaderSlackBot.java 后的输出结果，由于代码清单 3-5 中已经包含了全部内容，因此此处对其进行了截断。

代码清单 3-17　执行 ChannelReaderSlackBot.java 后的输出结果

```
Fatima [2023-08-11T09:04:20] : Hey everyone, I have an urgent
issue to discuss. I just got off a call with a client who's
experiencing app crashes as soon as they load it. They're
really frustrated. Can we get this sorted ASAP? :tired_face:

Keith [2023-08-11T09:04:35] : Thanks for bringing this to our
attention, Fatima. Let's jump on this right away. John, can you
take the lead in investigating the issue since our architect is
out sick today?

John [2023-08-11T09:04:52] : Sure thing, Keith. I'll dive into
```

```
the codebase and see if I can find any potential culprits for
the crashes.

John [2023-08-11T09:05:30] : Fatima, could you gather some
additional information from the client? Ask them about the
specific device, operating system, and any recent updates they
might have installed.

...
```

3.9　练习

因此，我们在这里显然还可以做一些额外的事情，这些步骤将留给你来完成，例如：

- 使 ChatGPTClient.java(相关类)对新手用户更加安全。例如，对于 ChatGPT，Top P 参数的有效值只能在 0 和 1 之间。如果用户指定的参数超出有效值范围，Chat.java 类的构造函数就会抛出异常。
- 将 ChannelReaderSlackBot.java 中读取 Slack 消息的代码连接到 ChatGPTClient.java 中，这样抓取消息和获取摘要就只需要一个步骤。
- 为 Slack Bot 本身添加更多功能，例如添加命令使频道中的任何人都能请求摘要。在当前状态下，机器人不会在频道中发布任何内容。但机器人的“用户接口”就是频道本身；因此，任何人都应能通过输入命令(如请求摘要)与 Slack Bot 交互。
- 确保机器人不会让糟糕的情况变得更糟。无论机器人何时提供摘要，它都不应该在频道中发布，因为这可能给本已嘈杂的情况增加更多噪声。最好让机器人向请求摘要的人发送一条私人消息(或你创建的任何新命令)。

3.10　小结

在本章中，我们讨论了人工智能在当今企业中的各种实际应用方式之一，并向你展示了如何通过使用构建器模式来改进 ChatGPTClient.java 应用程序，从而让类的构建比第 2 章更加灵活。

不过，最值得注意的是，我们讨论了什么是真正的“提示工程”，即提示工程不能仅通过向 ChatGPT 输入文本来实现。你肯定需要了解 ChatGPT API 所有输入参数的影响，才能正确、有效地执行提示工程。

利用所学到的提示工程知识，我们能够成功地获得所提供的任何大文本的摘要。最后，我们看到了运行自动机器人所需的代码，如果指定了有效的日期范围，该机器人就会以编程方式从任何 Slack 频道抓取消息。

在本章以及第 2 章中，我们只使用了 OpenAI API 的 Chat Completions 端点。在第 3 章中，将尝试使用 Whisper 和 DALL-E 端点，从而突破可能的极限。

第4章

多模态人工智能：用 Whisper 和 DALL-E 3 创建播客展示台

现在我们来介绍一个新名词：多模态人工智能。最简单地说，生成式人工智能模型可以创建以下格式的内容：

- 文本
- 音频
- 图片
- 视频

每一种格式都是一种模态。多模态人工智能是指使用多个人工智能模型共同生成(或理解)内容的过程，其中输入是一种模态，输出是另一种模态。

以 OpenAI 的 Whisper 模型为例。如果你向它提供音频，它就能将所说的一切转录成文本。同样的道理也适用于 DALL-E。如果你向它提供文字提示，它就能生成你所描述的图像。

在本章中，我们将把多模态人工智能提升到一个新水平！作为一名播客的忠实听众，我经常想，在听一个音频故事时，场景、图像、人物、主题或背景是什么样子的。

因此，我们将使用 OpenAI 的多模态模型创建一个播客展示台。虽然需要一些步骤，但最终效果令人惊叹。在收听播客讲述一个用豆腐烹饪美食的故事时(没试过就不要妄下断言)，播客展示台生成了图 4-1 所示的图像。

图 4-1 使用 GPT-4、Whisper 和 DALL-E 模型对有关豆腐的播客进行可视化的 AI 生成结果

为了让博客展示台的代码更容易理解，我们将分以下三个步骤分别执行：

- 第 1 步：获取一集播客，然后使用 Whisper 模型获取文字记录。
- 第 2 步：获取文字记录并使用 GTP-4 模型描述播客中讨论内容的视觉方面。
- 第 3 步：根据得到的描述使用 DALL-E 模型生成图像。

例如，本章介绍的代码就有大量实际用途：

- 如果只是对一集播客中的内容感到好奇(对我来说总是这样)，你可以获得一个简单的、有代表性的视觉图像，与你正在收听的内容联系起来。
- 对于有听力障碍的人来说，可以轻松地将播客或广播节目变成幻灯片式的图像。这大大提高了内容的可读性。

- 对于播客来说，现在有一种简单的方法，可以为每集节目添加视觉/主人公图像。这非常有用，因为 Apple 博客和 Spotify 等播客播放器使播客能够显示与单集相关的单张图片，这有助于提高听众的参与度。

4.1 介绍 OpenAI 的 Whisper 模型

现在让我们来介绍另一个新名词：**自动语音识别**(ASR)。普通消费者对这项技术非常熟悉，因为它已集成到手机(如 iPhone 的 Siri)和智能扬声器(如任何 Alexa 设备)中。ASR 技术的核心是将口语转换成文本。

Whisper 是 OpenAI 的语音识别模型，其准确率之高令人惊叹。代码清单 4-1 是备受欢迎的多邻国西班牙语播客的一集文字记录，该播客通过将英语和西班牙语结合在一起编成一个叙事故事，让英语听众也能轻松理解西班牙语。该文本使用 Whisper 模型生成。

代码清单 4-1 Whisper 模型执行语音识别，将音频转换为文本

```
...I'm Martina Castro. Every episode we bring you fascinating,
true stories to help you improve your Spanish listening and
gain new perspectives on the world. The storyteller will be
using intermediate Spanish and I'll be chiming in for context
in English. If you miss something, you can always skip back
and listen again. We also offer full transcripts at podcast.
duolingo.com.

Growing up, Linda was fascinated with her grandmother, Erlinda.
Erlinda was a healer or curandera, someone who administers
remedies for mental, emotional, physical, or spiritual
illnesses.

In Guatemala, this is a practice passed down orally through
generations in the same family. Mal de ojo, or the evil eye, is
considered an illness by many Guatemalans who believe humans
have the power to transfer bad energy to others. Neighbors
would bring their babies to Linda's grandmother when they
suspected an energy imbalance. Su madre lo llevaba a nuestra
casa para curarlo...
```

中文注释

……我是 Martina Castro。每集我们都会为你带来精彩的真实故事，帮助你提高西班牙语听力，并获得对世界全新的认识。讲故事的人将使用中级西班牙语，而我将用英语介绍故事的背景。如果你错过了什么，可以返回重听。我们还在 podcast.duolingo.com 上提供完整的文字稿。

在 Linda 的成长过程中，她非常爱祖母 Erlinda。Erlinda 是一名治疗师(curandera)，负责治疗心理、情绪、身体或精神疾病。

在危地马拉，这是一种家族世代口口相传的习俗。许多危地马拉人认为，"邪眼" (Mal de ojo)是一种疾病，他们相信人类有能力将不好的能量传递给他人。邻居们怀疑孩子能量失衡时，就会带孩子来找 Linda 的祖母。Linda 的祖母说："她的母亲会把孩子带到我们家来治疗……"。

如果你以前用过语音识别系统(即使是 Siri 和 Alexa 这样复杂的技术)，你就会知道此类系统也有问题，下面是一些例子。

- 语音识别在标点符号方面存在问题
  - 没有人会在说话时使用标点符号。在英语中，我们通过语调或音量的变化来提问或发出感叹，我们还用长短停顿来表示逗号和句号。
- 语音识别在处理外来词和口音时存在问题
  - 英语中至少有 17 万个单词(这要看你问的是谁)。然而，在英语会话中，我们总是使用外来词，下面列举几个示例。
    - Tsunami(源自日语)：通常由地震引起的巨大海浪
    - Hors d'oeuvre(源于法语)：开胃菜
    - Lingerie(源自法语)：女性内衣或睡衣
    - Aficionado(源自西班牙语)：热衷于某种特定活动或主题的人
    - Piñata (源自西班牙语)：颜色鲜艳的糖果盒，供孩子们不停地拍打
- 语音识别在人名方面存在问题
  - 某些人名、企业名和网站名通常难以拼写和理解
- 语音识别在处理同音字时存在问题
  - 你还记得那些发音相同但拼写和含义不同的单词吗？

本书的编辑对它们了如指掌！

· Would / Wood
· Flour / Flower
· Two Too To
· They're There Their
· Pair Pare Pear
· Break Brake
· Allowed Aloud

从代码清单 4-1 中可以看出，Whisper 能够理解音频中的所有标点符号，识别所有外来词(其中有几个)，并理解 URL 中的名称和公司名称(“duolingo”)！当然，如果你注意到的话，它还能听懂 wood 和 would 的区别。

4.2 Whisper 模型的特点和局限性

Whisper 模型能够将以下语言的口语音频转换成文本：

- 南非语
- 阿拉伯语
- 亚美尼亚文
- 阿塞拜疆语
- 白俄罗斯语
- 波斯尼亚文
- 保加利亚语
- 加泰罗尼亚语
- 汉语
- 克罗地亚语
- 捷克语
- 丹麦语
- 荷兰语
- 英语(当然！)

- 爱沙尼亚语
- 芬兰语
- 法语
- 加利西亚语
- 德语
- 希腊语
- 希伯来语
- 印地语
- 匈牙利语
- 冰岛语
- 印度尼西亚语
- 意大利语
- 日语
- 卡纳达语
- 哈萨克文
- 韩语
- 拉脱维亚文
- 立陶宛语
- 马其顿语
- 马来语
- 马拉地语
- 毛利语
- 尼泊尔语
- 挪威语
- 波斯语
- 波兰语
- 葡萄牙语
- 罗马尼亚文
- 俄语
- 塞尔维亚语

- 斯洛伐克语
- 斯洛文尼亚语
- 西班牙语
- 斯瓦希里语
- 瑞典语
- 他加禄语
- 泰米尔语
- 泰语
- 土耳其语
- 乌克兰语
- 乌尔都语
- 越南语
- 威尔士语

因此，最终它将能听懂你的音频，也可能听懂你的朋友和同事所说的任何语言。

开发人员每分钟只能向终端发送不超过 50 个请求，因此，如果你想转录大量音频，就必须考虑这一限制。

Whisper 支持 flac、mp3、mp4、mpeg、mpga、m4a、ogg、wav 或 webm 格式的音频。无论使用哪种格式，发送到终端的最大文件大小为 25MB。

现在，如果你没有广泛接触过音频文件，请注意有些格式会产生非常大的文件(如 wav 格式)，而有些格式则会产生非常小的文件(如 m4a 格式)。因此，将文件转换为其他格式可以帮助你解决 25 MB 的限制。不过，在本章稍后部分，我们将看到一个工具的代码，该工具可将一个大音频文件分成多个较小的文件。

4.3　转录终端

转录端点是一种 REST 服务，用于将音频转换为文本，仅与 Whisper 模型兼容。

4.3.1 创建请求

表 4-1 列出了调用转录端点需要的所有 HTTP 参数。

表 4-1 转录终点的 HTTP 参数

HTTP 参数	说明
端点 URL	https://api.openai.com/v1/audio/transcriptions
方法	POST
标题	Authorization：Bearer $OPENAI_API_KEY
内容类型	multipart/form-data

注意：

请密切注意上表中的内容类型！与以 JSON 对象形式发送所有 HTTP 请求参数的聊天端点不同，转录端点只接收作为**表单数据元素**的参数。如果尝试发送格式较好(甚至格式不好)的 JSON 对象，转录端点将返回一个非常模糊的错误。

4.3.2 请求正文(多部分表单数据)

表 4-2 列出了 Whisper 的请求体。

表 4-2 Whisper 的请求体

字段	类型	是否必需？	说明
File	file	必需	你希望转录的整个音频文件。 可接受的格式有 • flac • mp3 • mp4 • mpeg • mpga • m4a

(续表)

字段	类型	是否必需?	说明
File(续)			• ogg • wav • webm
Model	String	必需	要用于转录的模型的 ID。 兼容的模型包括 • whisper-1
prompt	String	可选	这是可提供的任何文本，用于更改模型的转录风格，或为其提供前一段音频的更多上下文。 请确保提示语言与音频相同，以获得最佳效果。 此外，该字段还可用于更改 Whisper 不熟悉的单词的拼写或大小写
response_format	String default:json	可选	这是转录输出的格式。 可接受的格式有 • json • text • srt • verbose_json • vtt
temperature	Number default: 0	可选	这是采样温度，范围从 0 到 1。 数值越大，输出的随机性越强；而数值越小，输出的确定性越强
language	String	可选	这是输入音频的语言。这是可选项，但提供该值可以提高转录的准确性并减少延迟

4.4 创建一个分割音频文件的实用程序：AudioSplitter.java

因此，我们几乎可以使用转录端点以编程方式调用 Whisper 模型了。不过，Whisper 模型对每个文件的限制是 25MB。

现在，如果你正在收听得克萨斯大学奥斯汀分校的 StarDate 播客，这不成问题，因为这个播客能让你在大约 2 分钟的音频中了解应该在夜空中寻找什么。

然而，其他音频节目却不是这样，它们往往长达一小时(甚至更长)。这种情况下，你的文件肯定会超过 25MB 的限制。

因此，让我们与 ChatGPT 结对编程，利用人类的智慧创建自己的工具，将单个音频文件分成多个较小的文件。

注意：

在本节中，我将介绍将大音频文件分成小块的一种可能性。例如，你可以使用流行的音频编辑应用程序(如开源工具 Audacity 或授权工具 Adobe Audition)手动将大文件分成小文件。

代码清单 4-2 是我发送给 ChatGPT 的提示，目的是获得一个分割音频文件的基本应用程序。

代码清单 4-2 提示。使用 ChatGPT 创建 AudioSplitter.java 应用程序

```
System: You are a Java developer
User: Write an application that takes as input a single MP3
file and splits the file into contiguous segments no longer
than 10 minutes. Name the file, "AudioSplitter.java"
```

中文注释

系统：你是一名 Java 开发人员。

用户：编写一个应用程序，将单个 MP3 文件作为输入，并将该文件分成不超过 10 分钟的连续片段。将文件命名为 AudioSplitter.java。

我创建了 AudioSplitter.Java 应用程序(如代码清单 4-3 所示)，其中包括我对 ChatGPT 生成内容的编辑。

代码清单 4-3　响应。AudioSplitter.java

```
import org.bytedeco.javacv.FFmpegFrameGrabber;
import org.bytedeco.javacv.FFmpegFrameRecorder;

import java.io.IOException;

public class AudioSplitter {
      public static void main(String[] args) {
      String inputFilePath = "path/to/file/sample.mp3";
      String outputDirectory = "path/to/folder/";
      int segmentDurationInSeconds = 600; // 10 minutes
      in seconds

      try (FFmpegFrameGrabber grabber = new FFmpegFrameGrabber
      (inputFilePath)) {
         grabber.start();

         long totalDurationInSeconds = (long) grabber.
         getLengthInTime() / 1000000; // Convert
         microseconds to seconds
         double frameRate = grabber.getFrameRate();

         long segmentStartTime = 0;
         long segmentEndTime;
         int segmentNumber = 1;

         while (segmentStartTime < totalDurationInSeconds) {
            String outputFilePath = outputDirectory +
            "segment_" + segmentNumber + ".mp3";
            try (FFmpegFrameRecorder recorder = new FFmpeg
            FrameRecorder(outputFilePath, 0)) {
               recorder.setAudioChannels(2);
               recorder.setAudioCodecName("libmp3lame");
               // Set the audio codec to MP3
               recorder.setAudioBitrate(192000); // Adjust
               bitrate as needed
               recorder.setSampleRate(44100); // Adjust
```

```
                sample rate as needed
                recorder.setFrameRate(frameRate);
                recorder.setFormat("mp3"); // Set the
                output format to MP3
                recorder.start();

                segmentEndTime = Math.min(segmentStartTime
                + segmentDurationInSeconds,
                totalDurationInSeconds);

                grabber.setTimestamp(segmentStartTime *
                1000000); // Set the grabber's timestamp to
                the start time in microseconds

                while (grabber.getTimestamp() / 1000000 <
                segmentEndTime) {
                   recorder.record(grabber.grabSamples());
                }
             }
             segmentStartTime = segmentEndTime;
             segmentNumber++;
          }
       } catch (IOException e) {
          e.printStackTrace();
       }
       }
}
```

由于标准 Java 库并不广泛支持各种媒体格式，因此我们结合使用了 FFmpeg 库和 JavaCV(两者都是免费开源的)。

目标很简单：使用 Java 语言将 MP3 文件分成不超过 10 分钟的连续片段。在这个简单的应用程序中，要完成以下步骤：

- 首先，指定输入文件路径、输出目录和所需的片段持续时间(以秒为单位)。
- 接下来，使用 FFmpegFrameGrabber 打开输入的 MP3 文件并收集相关信息，如帧速率、音频编解码器、采样率等。
- 然后，遍历输入的 MP3 文件，将其分割成指定时长(10 分钟或更短)的较小部分。对于每个片段，创建一个新的 FFmpegFrameRecorder，设置其参数，并在片段持续时间内录制帧。

- 最后，递增每个片段的片段开始时间和片段编号，直到处理完整个输入的 MP3 文件。

为了使该程序正常运行，你需要在项目中正确安装和配置 JavaCV 和 FFmpeg 库。

注意：

FFmpeg 是一个开源二进制文件，你需要在机器上安装，并将其放在 PATH 中或在项目中进行访问。JavaCV 通过 JNI(Java 本地接口)使用 FFmpeg。

FFmpeg 是一款用途极为广泛的媒体转换器，不仅能处理 MP3 音频文件，还能处理其他各种音频文件格式(包括 M4A、OGG 和 WAV)。它还能转换视频格式以及 PNG、JPEG 和 GIF 等静态图像。

在 MP3 文件上运行 AudioSplitter.java 实用程序后，你会得到一个文件夹，里面全是时长为十分钟或更短的分段音频文件。使用 AudioSplitter.java 实用程序，可以在一个 Java 文件中修改最适合你的设置。对于我们来说，这里的目标是获得小于 25MB 的音频文件，因此，如果要转录 8 小时的法律诉讼(例如 WAV 格式)，可能需要将持续时间调整得更短，如 6 分钟。

使用 AudioSplitter 时，最佳做法是将输出文件夹设置为与输入文件夹不同的文件夹，当我们开始使用 Whisper 模型调用转录端点时，你就会明白其中的原因。

4.5　创建音频转录器：WhisperClient.java

现在，让我们创建下一个 Java 应用程序 WhisperClient.java。我们将再次与 ChatGPT 结对编程，以获得工作基础。这一次，我们将要求在此应用程序中使用 OK HTTP 库，原因有以下两点。

- 第一，我们已经在第 3 章的 Slack Bot 应用程序中使用过该库。
- 第二，在使用 HTTP 多部分表单时，OK HTTP 库会让事情变得更简单一些。

代码清单 4-4 是在 Chat Playground 中开始工作时的提示。请务必注意，我要求HTTP请求超时60秒，因为Whisper可能需要一点时间来生成副本。

代码清单4-4 提示：要求ChatGPT将cURL转换为Java并发送到Whisper的API

```
System: You are a Java developer.
User: Convert the following code from cURL to Java, using
OkHttp to send the request. Make sure that I have a 60 second
timeout on my request. Iterate over a single folder on my
local computer and send all the files in the folder to the
webservice. Name the file, WhisperClient.java.

User: curl https://api.openai.com/v1/audio/transcriptions \
  -H "Authorization: Bearer $OPENAI_API_KEY" \
  -H "Content-Type: multipart/form-data" \
  -F file="@/path/to/file/audio.mp3" \
  -F model="whisper-1"

Model: gpt-4

Temperature: 1

Maximum Length: 1150
```

中文注释

系统：你是一名Java开发人员。

用户：将下面的代码从cURL转换为Java，使用OkHttp发送请求。确保请求有60秒的超时时间。遍历本地计算机上的单个文件夹，并将文件夹中的所有文件发送到Web服务。将文件命名为WhisperClient.java。

用户： curl https://api.openai.com/v1/audio/transcriptions \

-H "Authorization: Bearer $OPENAI_API_KEY"

-H "Content-Type: multipart/form-data"

-F file="@/path/to/file/audio.mp3" \

-F model="whisper-1"

模型：GPT-4

温度：1

最大长度：1150

经过一段时间后，ChatGPT给出了有效的响应，如代码清单4-5所示。

代码清单 4-5　响应：WhisperClient.java

```
import java.io.*;
import java.nio.file.*;
import okhttp3.*;
import java.util.*;
import java.util.concurrent.TimeUnit;
import java.util.stream.Collectors;
import java.util.stream.Stream;
/**
 * Client class to transcribe MP3 files using the OpenAI
   Whisper model.
 */
public class WhisperClient {

    public static void main(String[] args) throws IOException {
        // API key for OpenAI (this should be replaced with
           your actual API key)
        String openAIKey = "";
        // OpenAI transcription endpoint
        String endpoint = "https://api.openai.com/v1/audio/
        transcriptions";
        // Model used for transcription
        String model = "whisper-1";
        // Media type for the MP3 files
        MediaType MEDIA_TYPE_MP3 = MediaType.
        parse("audio/mpeg");
        // Folder containing the MP3 files to be transcribed
        String mp3FolderPath = "/Users/me/audio/segments";
        // Desired format for the transcription response
        String responseFormat = "text";

        // Configure the HTTP client with specified timeouts
        OkHttpClient client = new OkHttpClient.Builder()
            .connectTimeout(60, TimeUnit.SECONDS)
            .writeTimeout(60, TimeUnit.SECONDS)
            .readTimeout(60, TimeUnit.SECONDS)
            .build();

        // List to store all mp3 files from the directory
        List<File> mp3Files = new ArrayList<>();
```

```
// Try to collect all mp3 files in the directory and
    store them in the list
try (Stream<Path> paths = Files.walk(Paths.
get(mp3FolderPath))) {
    mp3Files = paths
        .filter(Files::isRegularFile)
        .filter(p -> p.toString().endsWith(".mp3"))
        .map(Path::toFile)
        . sorted(Comparator.comparing(File::getName)) //
        Sort the files alphabetically
        .collect(Collectors.toList());
} catch (IOException e) {
    System.out.println("File reading error: " +
    e.getMessage());
    return; // Exit if there's an error reading
    the files
}
// Iterate over each MP3 file, transcribe it, and print
    the response
for (File mp3File : mp3Files) {
    // Construct the request body for transcription
    RequestBody requestBody = new MultipartBody.
    Builder()
        .setType(MultipartBody.FORM)
        .addFormDataPart(
            "file",
            mp3File.getName(),
            RequestBody.create(mp3File, MEDIA_
            TYPE_MP3))
        .addFormDataPart("model", model)
        .addFormDataPart("response_format",
        responseFormat)
        .build();

    // Build the HTTP request
    Request request = new Request.Builder()
        .url(endpoint)
        .header("Authorization", "Bearer " + openAIKey)
        .post(requestBody)
        .build();

    // Make the request and process the response
    try (Response response = client.newCall(request).
```

```
            execute()) {
                if (!response.isSuccessful()) throw new
                IOException("Unexpected code " + response);
                System.out.println(response.body().string());
            } catch (IOException e) {
                System.out.println("Request error for file: " +
                mp3File.getName() + " - " + e.getMessage());
            }
        }
    }
}
```

正如你所看到的，OkHttpClient 类使我们能够使用构建器模式创建一个我们指定的超时的类实例。

使用 AudioSplitter.java 工具时，最好将输出文件夹与输入文件夹分开。我们正在创建一个目录内所有 MP3 文件的集合。更具体地说，它是一个名为 mp3Files 的 List<File>，用来保存我们要转录的所有文件。因此，如果要分成较小片段的文件与这些片段位于同一文件夹中，我们会将大文件(大于 25MB 的文件)连同较小的文件一起发送到转录端点，这就违背了使用 AudioSplitter.java 应用程序的初衷。

因此，Files.walk()方法使我们能够递归遍历 mp3FolderPath 目录，收集所有 MP3 文件，并过滤掉那些不是以.mp3 扩展名结尾的文件(为安全起见，并防止网络服务出错)。然后将每个“路径”映射到相应的 File 对象，并根据文件名按字母顺序排序。最后使用 Collectors.toList()方法将所有排序后的文件收集到 mp3Files 列表中。

有了 MP3 文件集，现在就可将它们发送到转录端点了。创建 RequestBody 时，最需要注意的几行是：

```
.addFormDataPart("model", model)
.addFormDataPart("response_format", responseFormat)
```

这是因为，如果你想在 HTTP 请求中添加任何可选参数(所有参数请参见表 4-2)，如 prompt 或 temperature，那么你需要在此处进行添加，就像我们指定模型和响应格式一样。

注意：

让我重申一遍——调用转录端点与聊天端点完全不同。你可能已经注意到 ChatGPTClient.java 和 WhisperClient.java 之间的主要区别之一是导入语句。ChatGPTClient.java(第 2 章和第 3 章)的导入语句中包含了 Jackson 库，因为我们需要将请求作为 JSON 对象发送。然而，WhisperClient.java 中的导入语句没有提及 Jackson，因为我们将以表单数据的形式发送所有内容。

4.6 用 Podcast 体验一下乐趣

好了，让我们用前面介绍的代码进行一次测试。This American Life 是 Ira Glass 主持的每周一次的公共广播节目(也是播客)，由芝加哥 WBEZ 合作制作，如图 4-2 所示。

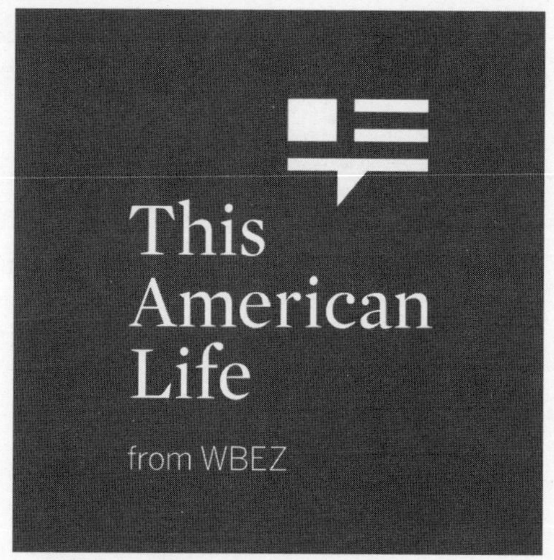

图 4-2 如果你正在寻找一个故事引人入胜的好播客，我推荐你收听 This American Life

每一集都围绕一个特定的主题或话题编排了一系列故事。有些故事是新闻调查，有些则是对普通人的采访，故事都很吸引人。第 811 集的标题是“我不能去的地方”，MP3 格式的文件大小为 56MB。鉴于 56MB 太大，无法发送到 Whisper 进行转录，因此代码清单 4-6 显示了 AudioSplitter.java 在 MP3 文件上的输出结果。

代码清单 4-6　在 This American Life 第 811 集上运行 AudioSplitter.java 的结果

```
[mp3 @ 0x139e9c6a0] Estimating duration from bitrate, this may
be inaccurate
Input #0, mp3, from '/Users/me/thislife/ep811.mp3':
  Metadata:
    encoder     : Lavf58.78.100
    comment     : preroll_1;postroll_1
  Duration: 00:58:58.34, start: 0.000000, bitrate: 128 kb/s
  Stream #0:0: Audio: mp3, 44100 Hz, stereo, fltp, 128 kb/s
Output #0, mp3, to '/Users/me/thislife/segments/segment_1.mp3':
  Metadata:
    TSSE        : Lavf60.3.100
  Stream #0:0: Audio: mp3, 44100 Hz, stereo, fltp, 192 kb/s
[libmp3lame @ 0x139ea92e0] 2 frames left in the queue
on closing
Output #0, mp3, to '/Users/me/thislife/segments/segment_2.mp3':
  Metadata:
    TSSE         : Lavf60.3.100
  Stream #0:0: Audio: mp3, 44100 Hz, stereo, fltp, 192 kb/s
[libmp3lame @ 0x13b167720] 2 frames left in the queue
on closing
Output #0, mp3, to '/Users/me/thislife/segments/segment_3.mp3':
  Metadata:
    TSSE         : Lavf60.3.100
  Stream #0:0: Audio: mp3, 44100 Hz, stereo, fltp, 192 kb/s
[libmp3lame @ 0x13b166df0] 2 frames left in the queue
on closing
Output #0, mp3, to '/Users/me/thislife/segments/segment_4.mp3':
  Metadata:
    TSSE         : Lavf60.3.100
  Stream #0:0: Audio: mp3, 44100 Hz, stereo, fltp, 192 kb/s
[libmp3lame @ 0x13b166df0] 2 frames left in the queue
on closing
Output #0, mp3, to '/Users/me/thislife/segments/segment_5.mp3':
  Metadata:
    TSSE         : Lavf60.3.100
  Stream #0:0: Audio: mp3, 44100 Hz, stereo, fltp, 192 kb/s
```

```
[libmp3lame @ 0x139ea35f0] 2 frames left in the queue
on closing
Output #0, mp3, to '/Users/me/thislife/segments/segment_6.mp3':
  Metadata:
    TSSE          : Lavf60.3.100
  Stream #0:0: Audio: mp3, 44100 Hz, stereo, fltp, 192 kb/s
[libmp3lame @ 0x139ea3540] 2 frames left in the queue
on closing
```

由于 AudioSplitter.java 工具使用 JNI 封装与 FFmpeg 接口，因此在音频分割过程中，你会看到很多诊断信息，如代码清单 4-6 所示。除非你关心编解码器、频率和比特率，否则大部分信息对你来说毫无意义。不过好消息是，我们现在有了一个包含 6 个 MP3 文件的文件夹，可以进行转录了！

当然，从代码清单 4-5 中的代码可以看到，WhisperClient.java 会遍历文件夹中的所有文件，并将它们发送到转录端点，以便使用 Whisper 模型。

代码清单 4-7 是一集完整副本的节选。

代码清单 4-7 This American Life 第 811 集的部分文字记录

```
"...My younger cousin Camille is not really a dog person, but
there is one dog she adored. Her name was Foxy, because she
looked exactly like a fox, except she was black. She was the
neighbor's dog, but she and Camille seemed to have a real
kinship, maybe because they both weren't very far from the
ground. Camille was around four or five years old back then,
and she had a little lisp, so Foxy came out as Fozzie. I
thought it was one of the cutest things I'd ever heard.

The way Camille remembers Foxy, it's almost like a movie. Her
memories feel like endless summer, hazy and perfect, like a
scene shot on crackly film. I just remembered like the feeling
of being excited to go and see Foxy. I have an image in my head
of like coming to the house, and I could see Foxy was like
outside. I can see Foxy through the door that leads to the
garden. There's a story about Camille and Foxy that I think
about fairly often. I've talked about it with my sister for
years, but never with Camille. And it's this. Once when they
were playing..."
```

中文注释

……我的小表妹 Camille 其实并不喜欢狗，但有一只狗她很喜欢。它的名字叫 Foxy，因为它长得和狐狸一模一样，只是它是黑色的。它是邻居家的狗，但它和 Camille 似乎真的很亲近，也许是因为它们都离地面不远。Camille 那时大约四五岁，说话有点口齿不清，所以将 Foxy 叫作 Fozzie。我觉得这是我听过的最可爱的事情之一。

Camille 回忆 Foxy 的方式就像一部电影。她的回忆就像无尽的夏天，朦胧而完美，就像用胶片拍摄的场景。我只是想起了那种去看 Foxy 的兴奋感。我脑子里浮现出一个画面：我来到家里，看到 Foxy 就在外面。我可以通过通往花园的门看到 Foxy。我经常想起 Camille 和 Foxy 的故事，我和我姐姐聊了很多年 Foxy，但从没和 Camille 聊过。事情就是这样。有一次，当她们在玩……

为简洁起见，我们只展示了文字记录的一部分。由于这一集的长度接近 1 小时，因此全文超过 8000 字。

4.7　走向 meta：提示工程 GPT-4 为 DALL-E 编写提示

由于要可视化播客情节的全文记录有数千字，我们将使用 GPT-4 自动创建 DALL-E 模型所需的提示。DALL-E 能接收提示中的文字描述并创建图像，但提示要尽量简短。代码清单 4-8 是 GPT-4 为 DALL-E 生成提示的提示。

代码清单 4-8　GPT-4 创建 DALL-E 的提示

```
System: You are a service that helps to visualize podcasts.
User: Read the following transcript from a podcast. Describe
for a visually impaired person the background and subject that
best represents the overall theme of the episode. Start with
any of the following phrases:
- "A photo of"
```

```
- "A painting of"
- "A macro 35mm photo of"
- "Digital art of "
User: Support for This American Life comes from Squarespace...
Model: gpt-4-32k
Temperature: 1.47
Maximum length: 150
Top P: 0
Frequency penalty: 0.33
Presence penalty: 0
```

中文注释

系统：你是一个帮助播客可视化的服务提供商。

用户：请阅读以下播客的文字记录。为视障人士描述最能体现该集整体主题的背景和主题。请从以下任一短语开始：

- 一幅照片
- 一幅画
- 微距 35mm 照片
- 数码艺术

用户：Squarespace 为《美国生活》提供支持……

模型：GPT-4-32k

温度：1.47

最大长度：150

Top P：0

频率惩罚：0.33

存在惩罚：0

正如你在提示中看到的，使用的模型是 GPT-4 的 32k 词元版本，便于我们处理超长的文本记录。DALL-E 需要知道要生成的图像类型，因此我们需要指定图像为照片、绘画、数字艺术等。需要确保 GPT-4 生成的文本是简短的，因此文本的最大长度为 150 个词元。此外，为防止 GPT-4 多次重复某些短语，我们引入了 0.33 的频率惩罚。

代码清单 4-9 显示了 GPT-4 在阅读《美国生活》(This American Life) 第 811 集的文字记录后得出的结果。

代码清单 4-9 GPT-4 创建的 DALL-E 提示

```
Digital art of a young girl sitting in a garden with a black
dog that looks like a fox. The girl is smiling and the dog is
wagging its tail. The image has a hazy, dream-like quality,
with crackly film effects to evoke nostalgia.
```

中文注释

一个小女孩和一只看起来像狐狸的黑狗坐在花园里的数字艺术。女孩面带微笑，小狗摇着尾巴。画面呈现出一种朦胧、梦幻的质感，裂缝般的胶片效果唤起人们的怀旧之情。

4.8 创建图像端点

要使用 DALL-E 模型从文本提示动态创建图像，需要调用“创建图像端点”。

4.8.1 创建请求

表 4-3 列出了调用创建图像端点需要的所有 HTTP 参数。

表 4-3 调用创建图像端点所需的 HTTP 参数

HTTP 参数	说明
端点 URL	https://api.openai.com/v1/images/generations
方法	POST
标题	Authorization：Bearer $OPENAI_API_KEY
内容类型	application/json

表 4-4 描述了创建图像端点请求体所需的 JSON 对象格式。由于明显的原因，prompt 是成功调用服务所需的唯一参数。

4.8.2 创建图像(JSON)

表 4-4 列出了创建图像端点的请求体。

表 4-4 创建图像端点的请求体

字段	类型	是否必需?	描述
prompt	String	必需	此处用于描述你希望创建的图像。 dall-e-2 的最大长度为 1000 个字符，dall-e-3 的最大长度为 4000 个字符
model	String	可选	生成图像的模型名称。 兼容的模型包括 • dall-e-2 • dall-e-3
n	Integer 或 null 默认值：1	可选	这是你要求创建的图像数量。 必须介于 1 和 10。 注意：由于 dall-e-3 所需的复杂性，OpenAI 可能将你的请求限制为单幅图像
quality	String 默认：standard	可选	用于指定要生成图像的质量。此参数仅适用于 dall-e-3。 接受的值为 • standard • hd
size	String 或 null 默认值： 1024x1024	可选	生成图像的大小。 dall-e-2 可用的图像尺寸为 • 256x256 • 512x512 • 1024x1024 dall-e-3 可用的图像大小为 • 1024x1024 • 1792x1024(横向) • 1024x1792(纵向)

(续表)

字段	类型	是否必需?	描述
style	String 默认值：vivid	可选	可用于指定生成图像的自然程度。 此参数仅适用于 dall-e-3。 可接受的值为 • natural(适合照片) • vivid(适合艺术效果)
response_format	String 或 null 默认值：url	可选	这是生成图像的格式。 可接受的值有 • url • b64_json
user	String	可选	这是代表最终用户的唯一标识符，可帮助 OpenAI 监控和检测滥用行为

4.8.3　处理响应

成功调用创建图像端点后，API 将响应一个图像 JSON 对象。表 4-5 是 Image(图像)对象的详细说明，该对象只有一个参数(或字段)。

表 4-5　Image(图像)JSON 对象的结构

字段	类型	说明
url 或 b64_json	String	如果请求中的响应格式为 url，则这是你所生成图片的 URL。 如果请求中的响应格式为 b64_json，则此字符串为经 base64 编码的 JSON 图像

4.9　创建图像生成器：DALLEClient.java

从表 4-4 和表 4-5 中可以看出，创建图像端点的行为与聊天端点非常相似：需要将为 DALL-E 模型指定的所有内容都封装在一个 JSON 对象中。因此，代码清单 4-10 中的代码也将使用 Jackson 库，因为我们将再次使用

JSON 对象。

代码清单 4-10　在 DALLEClient.java 中使用 Java 的 DALL-E API

```
import java.io.IOException;

import com.fasterxml.jackson.annotation.JsonProperty;
import com.fasterxml.jackson.databind.ObjectMapper;

import okhttp3.*;

public class DALLEClient {

    public static void main(String[] args) {
        String openAIKey = "";
        String endpoint = "https://api.openai.com/v1/images/
        generations";
        String contentType = "application/json";
        String prompt = "a 35mm macro photo of 3 cute
        rottweiler puppies with no collars laying down in
        a field";
        int numberOfImages = 2;
        String size = "1024x1024";

        OkHttpClient client = new OkHttpClient();
        MediaType mediaType = MediaType.get(contentType);

        // Create the Create Image JSON object
        CreateImage createImage = new CreateImage(prompt,
        numberOfImages, size);

        // Use Jackson ObjectMapper to convert the object to
            JSON string
        String json = "";
        try {
            ObjectMapper mapper = new ObjectMapper();
            json = mapper.writeValueAsString(createImage);
        } catch (Exception e) {
            e.printStackTrace();
            return;
        }
```

```
        RequestBody body = RequestBody.Companion.create(json,
        mediaType);
        Request request = new Request.Builder()
                .url(endpoint)
                .method("POST", body)
                .addHeader("Content-Type", contentType)
                .addHeader("Authorization", "Bearer " +
                openAIKey)
                .build();

        try {
            Response response = client.newCall(request).
            execute();
            if (!response.isSuccessful()) throw new
            IOException("Unexpected code " + response);
                System.out.println(response.body().string());
        } catch (Exception e) {
            e.printStackTrace();
        }
    }

    // Inner class for the CreateImage JSON Object
    public static class CreateImage {

        @JsonProperty("prompt")
        private String prompt;

        @JsonProperty("n")
        private int n;

        @JsonProperty("size")
        private String size;

        public CreateImage(String prompt, int n, String size) {
            this.prompt = prompt;
            this.n = n;
            this.size = size;
        }
    }
}
```

现在，由于我们已经使用 OkHttp 库通过 Whisper 模型的转录端点发出 HTTP 请求，我们将继续使用它来创建 DALL-E 模型的(Create Image)创建图像端点。

这里最需要了解的是 CreateImage 内部类。它具有@JsonProperty 注解，封装了创建图像所需的重要参数：

- 描述图片细节的文本提示
- 希望生成的图片数量
- 希望生成的图像大小

图 4-3 和图 4-4 显示了根据代码清单 4-9 中的文本提示生成的图像。

图 4-3　DALL-E 生成的 This American Life 播客第 811 集中女孩和小狗的图像(1)

图 4-4　DALL-E 生成的 This American Life 播客第 811 集中女孩和小狗的图像(2)

4.10　DALL-E 提示工程和最佳实践

现在，使用 DALL-E 创建图像需要利用提示工程，以获得一致、理想的结果。因此，我们建议你尝试使用不同的提示进行练习。也许你喜欢绘画而不是三维图像，也许你需要照片而不是数字艺术，也许你希望图片是特写镜头而不是肖像等，需要考虑的可能性很多。

无论你的使用情况如何，这里有两条黄金法则，可以让你从 DALL-E 提示中获得最大收益。

4.10.1　DALL-E 黄金法则 1：熟悉 DALL-E 可以生成的图像类型

首先，DALL-E 需要了解的最重要的一件事情就是需要生成的图像类型。下面列出了 DALL-E 能够生成的几种最常见图像类型：

- 三维渲染
- 绘画
- 抽象绘画
- 表现性油画
- 油画(任何已故艺术家的风格)
- 油画棒
- 数字艺术
- 照片
- 写实
- 超写实
- 霓虹灯照片
- 35 毫米微距照片
- 高品质照片
- 剪影
- 水汽
- 卡通

- 毛绒物体
- 大理石雕塑
- 手绘草图
- 海报
- 铅笔和水彩
- 合成波
- 漫画风格
- 手绘

4.10.2 DALL-E 黄金法则 2：描述你想要的前景和背景

为了得到一致、理想的结果，你需要对 DALL-E 进行描述，这一点再怎么强调也不够。这听起来可能有点奇怪，但向 DALL-E 描述图像的最佳方式就是像向另一个人描述梦境一样。

因此，作为你我之间的一个心理练习，请试着描述一下你最近的梦。当你描述梦中的人物、地点和事物时，你的脑海中就会浮现出你记得的最重要的事情，以及你所感受到的体验。再当你向另一个人描述时，一些细节开始浮现出来，比如：

- 当时有多少人在场(如果有)？
- 人或动物当时是什么姿势？站着、坐着还是躺着？
- 场景和背景中有哪些东西？
- 哪些物品让你印象深刻？声音？气味？颜色？
- 你有什么感觉？快乐、阴森、兴奋？
- 一天中感觉到的时间是什么？早晨、中午还是晚上？

如果你能向他人描述梦境，那么向 DALL-E 描述你想要的东西也应该不成问题。

4.11 小结

在本章中，我们完成了很多工作！通过几个类，我们创建了博客展示台。

- 首先，我们创建并使用了 AudioSplitter.java 类，它为我们提供了一个实用工具。如果你的音频文件大于 Whisper 模型的限制，该类将为你提供一个较小的音频文件的文件夹，以便发送到 Whisper。
- 其次，我们创建并使用了 WhisperClient.java 类来获取音频文件的文件夹的转录。文件夹可以包含单个音频文件或多个文件。唯一的限制是向转录端点和 Whisper 模型发送的请求数量。
- 接着，我们用 GPT-4 做了一点提示工程，以便根据转录内容获得播客中图像的描述性提示。
- 最后，我们创建并使用了 DALLEClient.java 类，以获取通过调用 GPT-4 模型生成的提示，并获得一张直观表示播客剧集的图像。

4.12 练习

除此之外，我们显然还可以做一些额外的事情，这些步骤将留给你来完成，例如：

- AudioSplitter.java 应用程序是 FFmpeg 的 Java 接口。FFmpeg 不仅能分割音频文件，还能对媒体文件进行更多处理，如格式转换和重新编码。做个实验，看看 Whisper 支持的媒体格式中，哪种是最小的音频文件。提示：肯定不是 WAV 格式。
- 如果你打算创建一个应用程序或服务，根据最终用户的文本提示自动生成图像，那么你肯定需要更新 DALLEClient.java 类，以确保你在 HTTP 请求中跟踪并提供用户参数。这是因为你的最终用户可能通过你的 API 密钥生成恶意图像。

请记住，你拥有 Open AI 的 API 账户，而用户没有！因此，你可能需要终止与使用你的服务但违反了 Open AI 内容规则的用户之间的业务关系。

第5章

使用 Discord 和 Java 创建自动社区管理器机器人

在发布应用程序或服务时，建立和维护自己的社区非常重要。以下是健康用户社区的标志：

- 成员参与有意义的讨论，分享见解、反馈和支持。
- 分歧或争论时有发生，但都得到了建设性的解决，不会诉诸人身攻击或贬损性语言。
- 这里有一种相互尊重的氛围，成员们相互倾听并承认不同意见的存在。
- 新老成员积极参与，确保社区保持活力，不会停滞不前。
- 用户贡献各种内容，从回答问题到分享资源，丰富了社区的知识库。
- 在给予和索取之间取得了平衡；寻求帮助或信息的成员也向他人提供帮助或信息。
- 新成员的加入很频繁，通常是由现有成员推荐的，这表明社区得到了积极评价，值得推荐。
- 用户通常会成为社区或平台的拥护者，在直接的社区空间之外(如社交媒体或其他论坛)宣传社区或平台。

- 社区通过为应用程序或服务的特性和功能提供新创意，帮助塑造应用程序或服务。

无论我创建什么类型的应用程序或服务，都希望用户社区具有以上特点！

5.1 选择 Discord 作为社区平台

在过去几年中，Discord 作为一种有用的社区管理工具受到对社区充满热情的人们的青睐。这部分归功于它的跨平台兼容性，无论成员使用的是台式机、移动设备还是网页浏览器，都能保持联系。不过，它的一个突出特点是基于邀请的社区系统，可以帮助社区管理者控制增长并防止垃圾邮件。这种模式不仅能确保为成员提供量身定制的体验，还能增强安全性，因为社区管理者可以自行决定是否授予访问权限。

Discord 不仅支持文本消息，还支持语音聊天和流媒体视频。Discord 与 Slack 非常相似，社区经理可将内容分成不同的频道，以组织讨论、简化信息流，并帮助用户查看他们感兴趣的内容。

5.2 创建比 Slack 机器人更高级的机器人

现在，如果你已经成功地完成了第 3 章中使用 Slack 机器人的步骤，那么本章的步骤对你来说就再熟悉不过了。在第 3 章中，我们创建了一个 Slack 机器人，用于在一段时间内阅读单个频道，并获取讨论内容的摘要。Slack 机器人并不是社区经理，更像是一个有用的助手。

在本书的剩余部分中，我们将执行所有必要步骤，为 Discord 制作强大的机器人，利用人工智能帮助实际管理社区。

5.3　创建比普通 Discord 机器人更高级的机器人

如果你具有使用 Discord 机器人的经验，那么你可能会知道与它们交互的最常见方式就是所谓的“/command”。这使得典型的机器人(非智能机器人)基本上只能在收到非常具体的操作或命令时才能工作。如果不提供“/ command”，机器人就会保持沉默，什么也不做。从根本上说，它体现了“有话才说”的特点。

不过，我们要创建的 Discord 机器人将是人工智能机器人，因此它将比一般的 Discord 机器人先进得多。我们要创建的机器人将能够阅读和查看 Discord 服务器中的所有消息，并足够智能地做出正确的回应。

了解机器人的角色

让我们来探讨一个比较真实的场景。我们正在创建一个公共 Discord 服务器，用于与移动银行应用程序的用户进行交互。我们的最终目标是使用 Java 编写的机器人来处理以下场景。

- 问答：监控特定频道，自动回答用户关于如何使用银行应用程序的问题。为此，需要对机器人进行培训，使其了解应用程序的工作原理。
- 不招揽顾客：对于任何商业社区而言，重要的是社区参与者不会成为不法分子的目标。例如，如果你正在创建一个银行应用程序，你希望用户名为 B4nk Admin 的人联系你的客户吗？

注意，对于任何社区而言，保护成员免受仇恨语言等有害内容的侵害都是非常重要的。

5.4　银行示例：克鲁克银行

在本示例中，我决定使用一个虚构的银行名称，该名称与真实银行名称重合的可能性极低。因此，在本例中，“克鲁克银行”将为其客户推出一款新的移动应用程序，如图 5-1 所示。他们希望有一个由机器人监控的频道来回答应用程序用户的问题，同时希望确保没有人在他们的 Discord 服务器中招揽用户，或发布具有伤害性或有害的内容。

图 5-1 这家虚拟银行的虚拟应用程序存在实际问题

5.5 第一件事：创建自己的 Discord 服务器

在制作人工智能 Discord 机器人之前，我们显然需要一个已经就位的 Discord 服务器供机器人交互使用。使用 Discord App 或访问 Discord 网站(当然要先登录)，然后启动添加/创建新服务器的流程。

启动流程后，选择 Create My Own 选项，如图 5-2 所示。

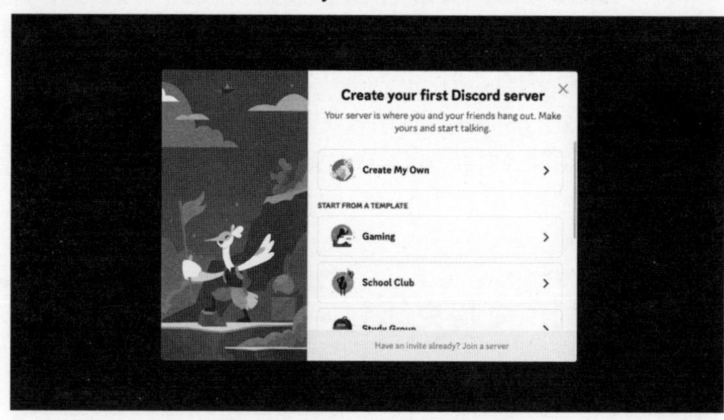

图 5-2 创建自己的 Discord 服务器

接下来，系统会提示你指定有关服务器的其他信息。如图 5-3 所示，继续完成创建过程，直到提示你提供服务器的名称和图标。

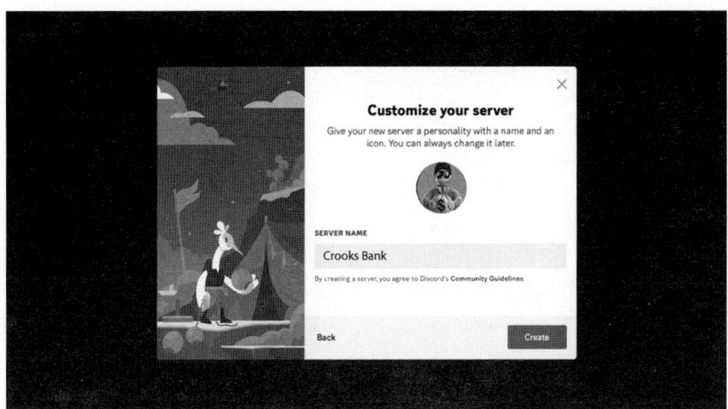

图 5-3　为自己的 Discord 服务器提供名称

指定服务器名称并提供可选的服务器图标(如果有)。

5.6　创建问答频道

默认情况下，每个 Discord 服务器都有一个“通用”频道，但我们希望有一个专门的问答频道。根据你创建服务器的方式，图 5-4 和图 5-5 将显示创建新频道的信息。

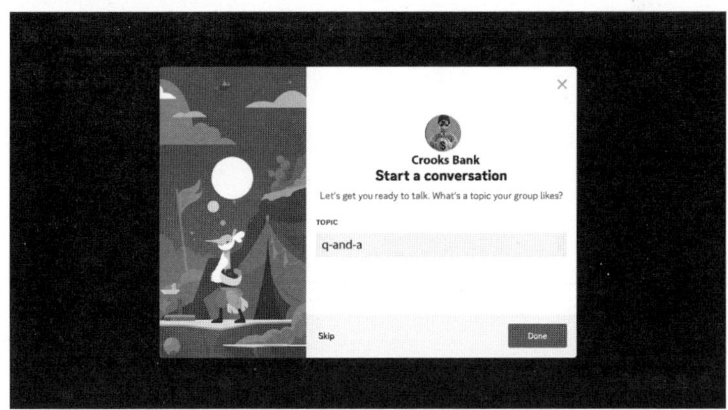

图 5-4　使用 Web 界面创建频道

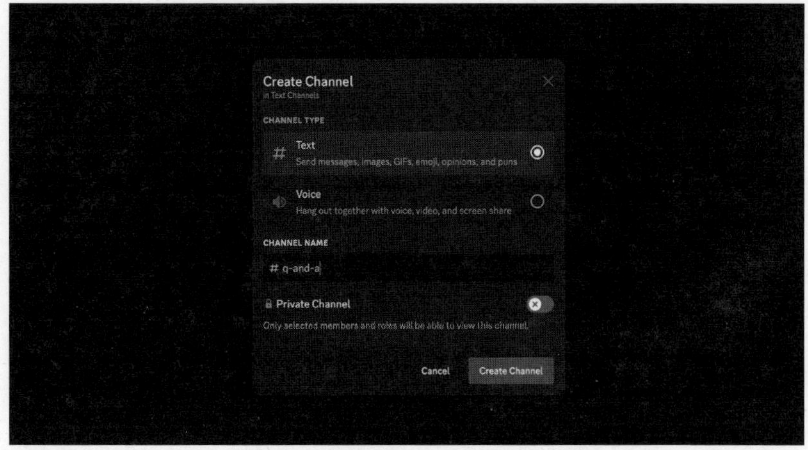

图 5-5　使用 Discord 应用程序创建频道

5.7　使用 Discord 注册新的 Discord 机器人应用程序

既然 Discord 服务器已经创建了相应的频道，那么现在就该注册机器人本身了——或者说，在我们的例子中是机器人本身。为了保持代码的简洁性和可管理性，将为 Discord 服务器安装多个机器人。第一个机器人将专门用于“问答”频道中的问题。第二个机器人将监控所有频道，防止有害内容或招揽生意等不受欢迎的内容。

要创建机器人，请访问 Discord 开发人员网站：

https://discord.com/developers

在页面右上方，单击 New Application 按钮，如图 5-6 所示。

在 Discord 和 Slack 术语中，“机器人”就是“应用程序”，除非先在 Discord 注册，否则不允许机器人在 Discord 服务器上运行。

为机器人指定一个名称，然后单击 Create 按钮，如图 5-7 所示。

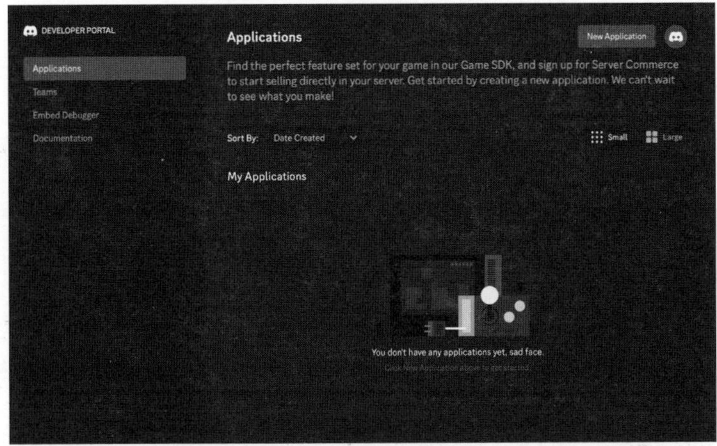

图 5-6　要创建 Discord 机器人，请访问 Discord 开发人员网站

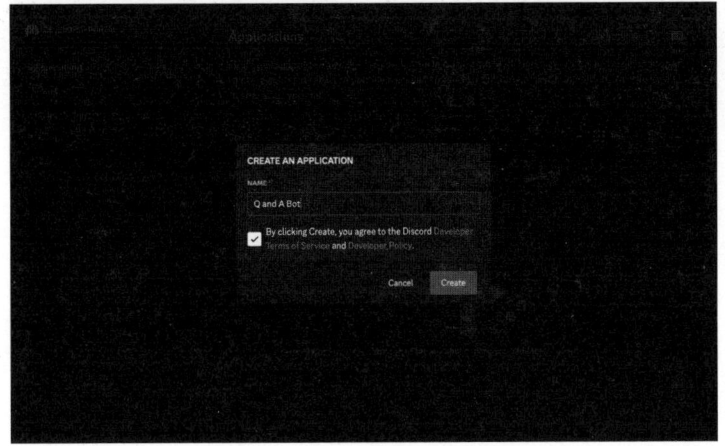

图 5-7　为 Discord 创建/注册机器人

5.8　指定机器人的基本信息

随后，你将进入一个页面，在这里可以指定机器人的一般信息，如图 5-8 所示。

请务必熟悉页面左侧的导航菜单。如你所见，我们有几类设置可供机器人配置。默认情况下，我们进入的是 General Information 页面，在这里可以指定机器人的基本信息。如果你已经为机器人准备好了图标，可在此处上传。

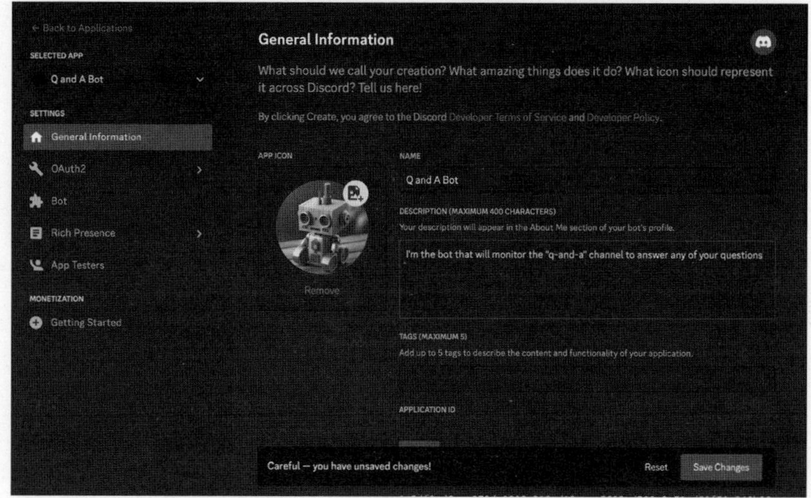

图 5-8　我决定给机器人上传一个可爱的小机器人图标

5.9　为机器人指定 OAuth2 参数

现在是为机器人指定作用域和权限的时候了。如果你按照第 3 章中创建 Slack 机器人的步骤执行过操作，那么你会对这个步骤感到熟悉。机器人不能也不应该拥有无所不能的能力，它们只能执行设计时所列出的一系列操作。

在左侧的设置导航菜单中，导航至 OAuth2 | URL Generator 以继续下一步。

下面是我们需要的作用域：

- 作用域(SCOPES)
    - 机器人(bot)

如图 5-9 所示。

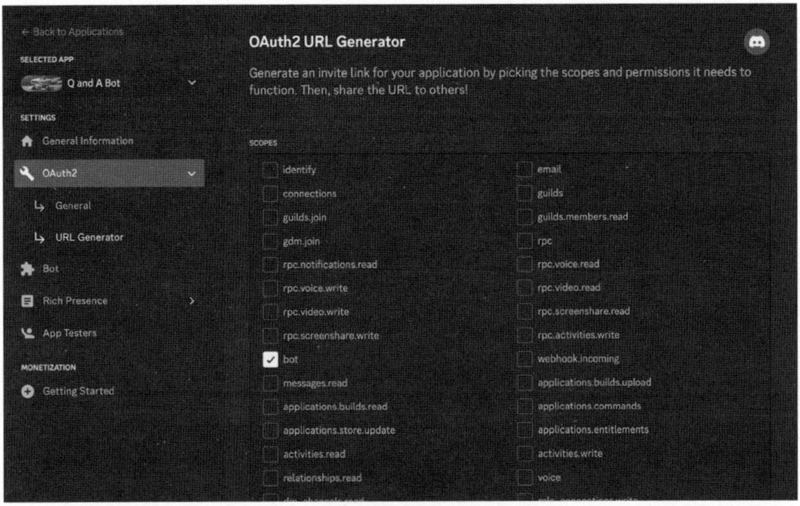

图 5-9　选择作用域

选择机器人的作用域后，就可以看到只适用于机器人的所有权限。机器人权限分为三类：常规、文本和语音。

如果你对这些类别感到好奇，一般权限允许机器人以普通人类审核者的身份行事，例如管理服务器、角色和频道。拥有这些权限的机器人还可以封禁成员。

文本权限允许机器人在文本频道收发信息，语音权限允许机器人参与语音频道。很简单吧？

为机器人选择以下权限：

- 机器人权限(BOT PERMISSIONS)
    - 文本权限(TEXT PERMISSIONS)
        - 发送消息(Send Messages)
        - 读取消息历史记录(Read Message History)

相应权限如图 5-10 所示。

虽然你还没有编写任何 Java 代码，但现在是时候邀请你的机器人加入服务器了。

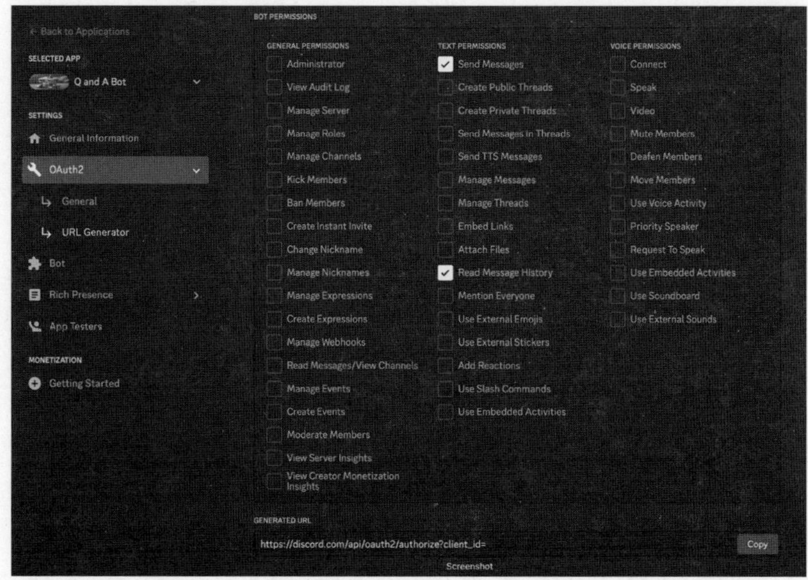

图 5-10　选择文本权限

5.10　邀请机器人加入服务器

如图 5-10 所示，选择了适当权限后，Discord 会显示一个动态生成的 URL，让你可以邀请机器人加入服务器。

复制该 URL 并将其粘贴到已验证过 Discord 的 Web 浏览器中。结果如图 5-11 所示。

单击 Continue 按钮将机器人加入服务器。

接下来，你会看到一个与前一页非常相似的页面(如图 5-12 所示)，但主要区别是它会列出关于机器人所有权限和功能的摘要。通常情况下，如果你要将机器人添加到一个你没有创建的服务器上，这个页面将非常有用。不过，因为这个机器人是我们自己创建的，所以这只是为了确认我们之前已经指定的设置。

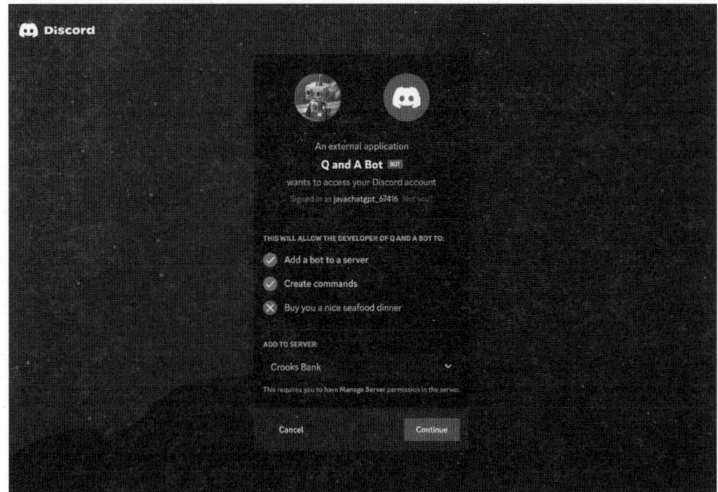

图 5-11　如果你仔细阅读该页面，就会发现 Discord 很有幽默感

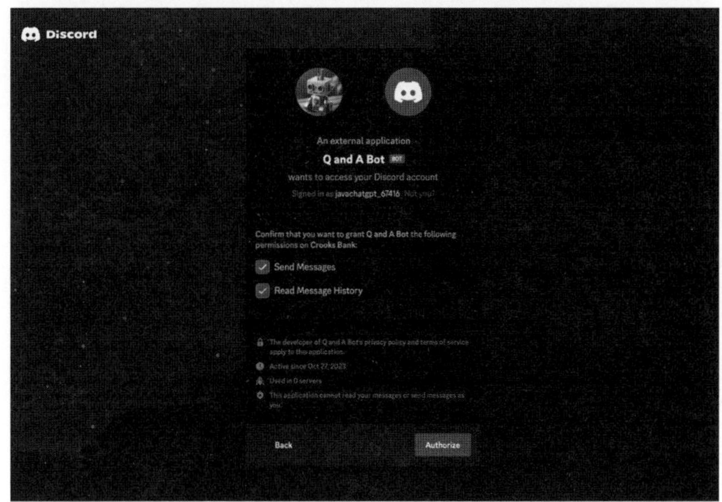

图 5-12　确认机器人的功能

单击 Authorize 按钮，授予机器人在服务器上运行的权限。

如果一切进行顺利，你就会在服务器的常规频道中看到一条自动消息，说明授权过程已经成功。

5.11 为机器人获取 Discord ID 词元并设置网关 Intent

现在是为你的机器人获取 Discord ID 词元(token)的时候了，你将在代码中用它以编程方式验证你的机器人。

注意：

显而易见，在此处使用“词元”一词让我很紧张，因为根据上下文，这个词在本书中有两个不同的含义，这里要快速复习一下这两个词的含义：

- 在使用 Discord 和 Slack API 时，“词元”就是身份验证词元。
- 在使用 OpenAI API 时，“词元”是单词的一部分。

返回 Discord 开发人员网站，单击设置导航菜单中的 Bot 类别继续下一步。

虽然你还没有看到词元，但你需要单击 Reset Token 按钮，如图 5-13 所示。

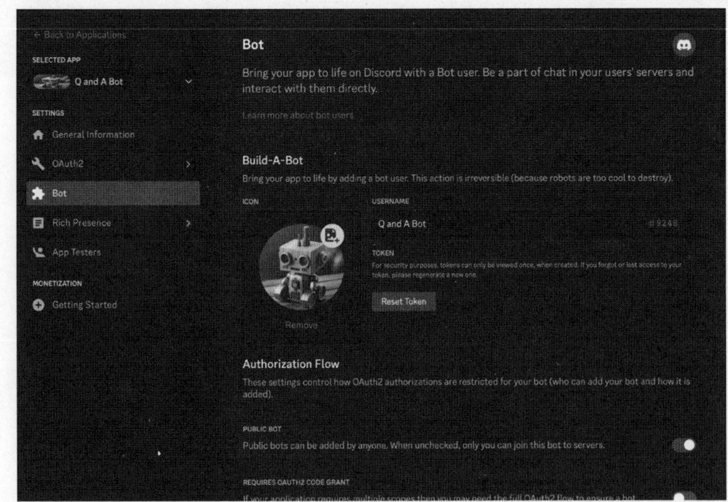

图 5-13 单击 Reset Token 按钮查看 ID 词元

务必将 ID 词元复制并保存到安全的地方，因为在本章后面介绍的 Java 代码中，你将需要这个词元。

向下滚动页面至 Privileged Gateway Intents 部分，启用 MESSAGE CONTENT INTENT 选项。

注意：

让我们放缓节奏，来谈谈 Intent。究竟什么是 Intent，为什么需要它？就 Discord API 而言，你需要明确指定你希望 Discord 以编程方式通知的每种类型的信息。否则，Discord 会不断向你发送与你或你的机器人无关的事件。例如，就我们而言，我们并不关心用户何时加入或离开服务器。但是，如果你想向任何首次加入服务器的人发送服务器规则列表，那么你肯定希望启用 SERVER MEMBERS INTENT。当深入研究代码时，你会看到更多关于 Intent 的信息。

请务必单击绿色按钮 Save Changes 以保存更改设置。结果如图 5-14 所示。

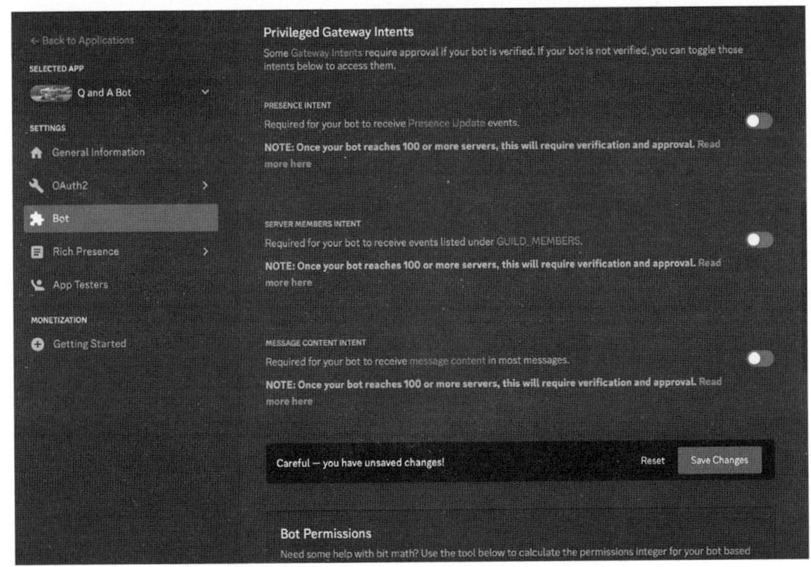

图 5-14　启用 MESSAGE CONTENT INTENT 选项

5.12 用 Java 创建问答机器人应用程序，回答来自频道的问题

我们已经完成了所有必要的先决条件，也知道了用于监控用户提问的频道名称，现在让我们用 Java 编写代码，加入服务器并访问特定 Discord 频道的所有消息。

设置依赖关系

Java Discord API(JDA)库为开发人员提供了一种非常直接的方法来创建自动应用程序，以便与 Discord 服务器一起工作。我们需要的大部分内容都来自 net.dv8tion.jda.api 包，它存在于 net.dv8tion-<VERSION> jar 文件中。

JDA 库包含如下的依赖项：

- Java 注释 API(增加了对基本注释的支持)
  - javax.annotation-api-<VERSION>.jar
- Opus Java(用于实时音频通信的 Java 库)
  - opus-java-<VERSION>.jar
- Neovisionaries websocket 客户端(通过 Web 套接字进行通信的 Java 库)
  - nv-websocket-client-<VERSION>.jar
- OK HTTP(我们已经熟悉这个用于 HTTP 通信的库)
  - okhttp-<VERSION>.jar
- Apache Commons(Java 开发人员常用的库)
  - commons-collections4-<VERSION>.jar
- SLF4J(我们已经熟悉这个用于记录日志的库)
  - slf4j-api-<VERSION>.jar

5.13　创建第一个 Discord 机器人：TechSupportBotDumb.java

这是本章要创建的两个 Discord 机器人中的第一个。这个名为 TechSupportBotDumb.java 的机器人将负责观察 Discord 服务器中问答频道的消息。

稍后将创建另一个机器人，它将负责管理 Discord 服务器中的所有内容，包括“问答”频道中的不受欢迎内容。这里的目标是遵循“关注点分离”的架构模式。我们不会创建一个大型 Java Discord 机器人来满足 Discord 服务器的所有审核需求，而是将这些功能分配到两个不同的应用程序中。

还将循序渐进地介绍 Java Discord 功能的学习曲线。在本书的最后一章，将增强这两个机器人的功能，并使用开放式人工智能 API 使其成为人工智能机器人。

代码清单 5-1 是我们创建一个基本 Discord 机器人所需的代码，该机器人可以查看单个频道中发布的所有消息并提供回复。

代码清单 5-1　TechSupportBotDumb.java

```
import java.io.IOException;
import java.util.EnumSet;

import net.dv8tion.jda.api.JDA;
import net.dv8tion.jda.api.JDABuilder;
import net.dv8tion.jda.api.entities.Activity;
import net.dv8tion.jda.api.entities.User;
import net.dv8tion.jda.api.entities.channel.ChannelType;
import net.dv8tion.jda.api.events.message.MessageReceivedEvent;
import net.dv8tion.jda.api.hooks.ListenerAdapter;
import net.dv8tion.jda.api.requests.GatewayIntent;

// This class extends a ListenerAdapter to handle message
events on Discord.
public class TechSupportBotDumb extends ListenerAdapter {

    // The bot's Discord token for authentication.
    static String discordToken = "YOUR_DISCORD_BOT_TOKEN";
    // The name of the channel the bot should monitor and
```

```
interact with.
static String channelToWatch = "q-and-a";

public static void main(String[] args) throws IOException {

    // Set of intents declaring which types of events the
    bot intends to listen to.
    EnumSet<GatewayIntent> intents = EnumSet.of(
            GatewayIntent.GUILD_MESSAGES, // For messages
            in guilds.
            GatewayIntent.DIRECT_MESSAGES, // For private
            direct messages.
            GatewayIntent.MESSAGE_CONTENT // To allow
            access to message content.
    );

    // Initialize the bot with minimal configuration and
    the specified intents.
    try {
        JDA jda = JDABuilder.createLight(discordToken,
        intents)
                .addEventListeners
                (new TechSupportBotDumb()) // Adding the
                current class as an event listener.
                .setActivity(Activity.customStatus
                ("Ready to answer questions")) // Set the
                bot's custom status.
                .build();

        // Asynchronously get REST ping from Discord API
        and print it.
         jda.getRestPing().queue(ping -> System.out.
         println("Logged in with ping: " + ping) );
        // Block the main thread until JDA is fully loaded.
        jda.awaitReady();

        // Print the number of guilds the bot is
        connected to.
        System.out.println("Guilds: " + jda.
        getGuildCache().size());
    } catch (InterruptedException e) {
        // Handle exceptions if the thread is interrupted
        during the awaitReady process.
        e.printStackTrace();
    }
```

```
    }

    // This method handles incoming messages.
    @Override
    public void onMessageReceived(MessageReceivedEvent
    messageEvent) {
        // The ID of the sender.
        User senderDiscordID = messageEvent.getAuthor();

        // Ignore messages sent by the bot to prevent selfresponses.
        if (senderDiscordID.equals(messageEvent.getJDA().
        getSelfUser())) {
            return;
        } else if (messageEvent.getChannelType() ==
        ChannelType.TEXT) {
            // Ignore messages not in the specified "q-and-a"
            channel.
            if (!messageEvent.getChannel().getName().equalsIgno
            reCase(channelToWatch)) {
                return;
            }
        }
        // Send a greeting response to the user who sent the
        message.
        String reply = "hi <@" + senderDiscordID.getId() + ">,
        I can help you with that!";
        messageEvent.getChannel().sendMessage(reply).queue();
    }
}
```

在我们的类中，需要扩展 JDA API 中的 ListenerAdapter 类，这样才能正常工作。现在，当你分析 TechSupportBotDumb.java 时，会发现我们把事情变得非常简单，因此只需要关注两个方法：main() 和 onMessage-Received()。

在类的开头，你还会注意到我们用 channelToWatch 变量指定了要监控的频道。

注意：

由于某些原因，Discord 自己的术语有时将 Discord 服务器称为 guild(公会)。因此，JDA 库在提及 Discord 服务器时也会使用 guild 一词。不过，从我们的角度看，guild 就是 Discord 服务器。

在 main()方法中，有一个 GatewayIntent 集合(具体来说，它是一个 EnumSet，但归根结底是一个集合)。大家可能还记得，使用 Intent 可以明确指定感兴趣的信息类型。在我们的例子中，我们感兴趣的是：

- 发送到服务器的信息(guild 信息)
- 用户直接发送给机器人的消息(直接消息)
- 发送的消息内容(消息内容)

之后，我们再次使用 JDABuilder 类的构建器模式，并调用以下 discordToken 和我们感兴趣的意图：

```
JDA jda = JDABuilder.createLight(discordToken, intents)
```

5.13.1 喜欢使用 Lambda 表达式来简化代码

在 main()方法中，我们通过 Lambda 表达式使用 JDA 库异步向 Discord 服务器发送 ping 请求。与所有网络请求一样，如果不是以异步方式完成，那么主线程将被阻塞，直到收到响应为止，这是一件坏事。因此，在收到 ping 响应后，我们执行 println()语句来显示 ping 请求到达服务器所需的时间。使用 Java Lambda 表达式的代码如下。

```
jda.getRestPing().queue(ping -> System.out.println("Logged in
with ping: " + ping) );
```

那么，如果不使用 Lambda 表达式来获取与 Discord 服务器的 ping 时间呢？代码看起来会更像这样：

```
// instantiate a new PingConsumer
jda.getRestPing().queue(new PingConsumer());
...
// define the PingConsumer as an inner class
class PingConsumer implements Consumer<Long> {
```

```
        @Override
        public void accept(Long ping) {
            System.out.println("Logged in with ping: " + ping);
        }
    }
```

如果不使用 Lambda，我们就需要创建一个实现 Consumer 接口的内部类(老实说，我们并不关心这个接口)。通过实现该接口，我们需要实现 accept()方法，该方法将在响应返回时异步调用。然后，我们将在 jda.getRestPing().queue()方法调用中创建一个新的 PingConsumer 实例。

5.13.2　处理发送到 Discord 服务器的消息

对第一个 Java Discord 机器人进行封装时，我们需要谈谈 onMessageReceived()方法。Discord 服务器发送的每一条消息，以及用户以 DM 形式直接发送给机器人本身的消息，都会异步调用该方法。

注意：

你知道吗，当机器人在 Discord 服务器中发送对用户问题的回答时，Discord 会调用机器人的 onMessageReceived()方法，将机器人刚刚发送的消息发送给机器人。这听起来像是无限循环，不是吗？因此，我们设置了逻辑，让机器人忽略自己发送的消息。

在 onMessageReceived()方法的最后几行，我们通过在回复中使用“@标记”，向信息的原始发送者呈现一个友好的响应。正如之前提到的，第一版问答机器人是“笨”的。在 Discord 服务器上发布问题时，它会对问题做出响应，但并不会真正回答你的问题。

5.13.3　成功！运行你的第一个 Discord 机器人：TechSupportBotDumb.java

现在运行 Java Discord 机器人。执行应用程序后，务必返回 Discord 服务器，并尝试在你设置的问答频道中输入问题。图 5-15 是对“这个机器人会回答我关于应用程序的问题吗？”的响应。

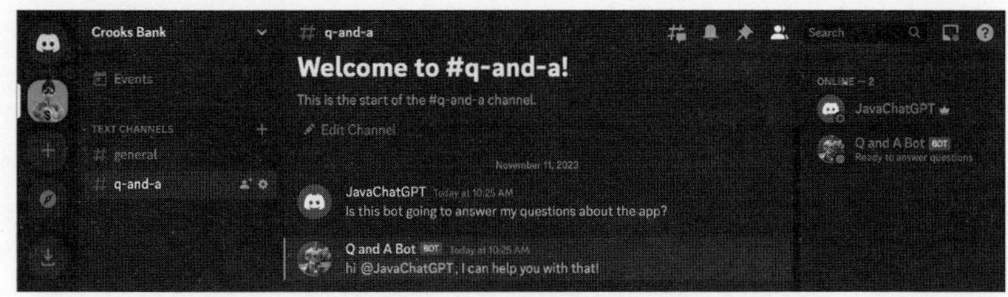

图 5-15 在 Discord 中运行

仔细观察图 5-15，你会看到一些关键功能，例如：

- 在右侧，你会看到机器人在线，状态指示灯为绿色。
- 机器人还有一个自定义状态，显示它将在频道中做些什么。
- 在频道中提问后，机器人会直接进行标记。

5.14 简化在 Discord 注册下一个 Discord 机器人应用程序的流程

既然我们已经成功执行了所有步骤，获得了一个正常运行的 Discord 机器人，那么创建第二个机器人将是小菜一碟！只需要简要重申一下创建第二个 Discord 机器人的所有步骤。由于第二个机器人将充当审核者，而不是为 Discord 服务器的用户提供问题解答，我将确保指出需要更改或增强的项目。

5.14.1 在 Discord 注册新的 Discord Bot 应用程序

执行与上述相同的步骤；不过，最好给机器人取一个不同的名字。可将第二个机器人命名为“内容审核机器人”。

5.14.2 指定机器人的一般信息

我为内容审核机器人设置了一个不同的图标，因此这里进行了指定(图 5-16)。

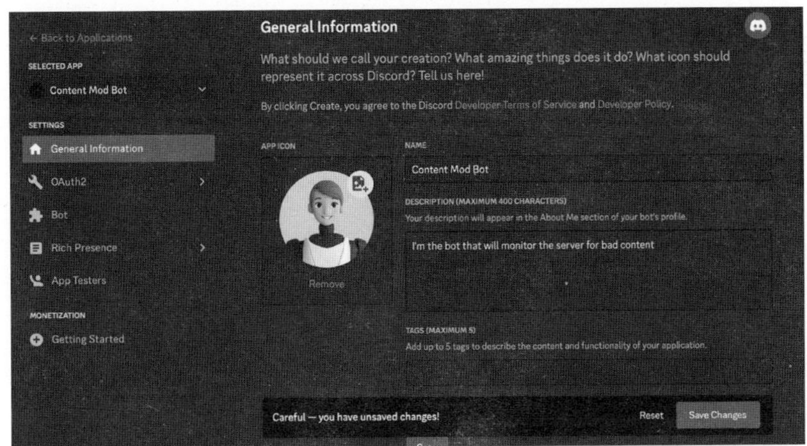

图 5-16　为第二个机器人提供名称和图标

5.14.3　为机器人指定 OAuth2 参数

第二个机器人需要更多权限才能执行更多任务。下面是需要的作用域：

- 作用域
  - 机器人

为机器人选择以下权限：

- 机器人权限
  - 一般权限
    - 取消会员
    - 封禁会员
  - 文本权限
    - 发送消息
    - 管理消息
    - 阅读消息历史

5.14.4　将机器人加入服务器

对第一个机器人重复上述步骤。

5.14.5 为机器人获取 Discord ID 词元并设置网关 Intent

再次按照上述步骤获取 Discord ID 词元。然后向下滚动页面，到达 Privileged Gateway Intents 部分，启用 SERVER MEMBERS INTENT 和 MESSAGE CONTENT INTENT 选项。

5.15 创建下一个 Discord 机器人：ContentModeratorBotDumb.java

内容管理器的作用是确保不在 Discord 服务器上发布多余的内容。就像我们在本章前面创建的上一个机器人一样，这个机器人目前还不具备人工智能。在当前状态下，机器人会不加区分地删除服务器上任何地方发布的包含“小狗”一词的信息。

这并非因为小狗天生邪恶。不过，如果不加管束，它们确实会毁掉你最喜欢的一双鞋。老实说，我们只是需要在运行机器人时测试 Discord 中的代码。

代码清单 5-2 显示了 ContentModeratorBotDumb.java 的内容。

代码清单 5-2 ContentModeratorBotDumb.java

```
import java.io.IOException;
import java.util.EnumSet;

import net.dv8tion.jda.api.JDA;
import net.dv8tion.jda.api.JDABuilder;
import net.dv8tion.jda.api.entities.Activity;
import net.dv8tion.jda.api.entities.Message;
import net.dv8tion.jda.api.entities.User;
import net.dv8tion.jda.api.entities.channel.unions.
MessageChannelUnion;
import net.dv8tion.jda.api.events.message.MessageReceivedEvent;
import net.dv8tion.jda.api.hooks.ListenerAdapter;
import net.dv8tion.jda.api.requests.GatewayIntent;

// This class extends a ListenerAdapter to handle message
```

```
events on Discord.
public class ContentModeratorBotDumb extends ListenerAdapter {

    // The bot's Discord token for authentication.
    static String discordToken = "YOUR_DISCORD_BOT_TOKEN";
    static String bannedWord = "puppies";

    public static void main(String[] args) throws IOException {
    // Set of intents declaring which types of events the
    bot intends to listen to.
    EnumSet<GatewayIntent> intents = EnumSet.of(
            GatewayIntent.GUILD_MEMBERS, // to get access
            to the members of the Discord server
            GatewayIntent.GUILD_MODERATION, // to ban and
            unban members
            GatewayIntent.GUILD_MESSAGES, // For messages
            in guilds
            GatewayIntent.MESSAGE_CONTENT // To allow
        access to message content
    );

    // Initialize the bot with minimal configuration and
    the specified intents.
    try {
        JDA jda = JDABuilder.createLight(discordToken,
        intents)
                .addEventListeners(new
                ContentModeratorBotDumb()) // Adding the
                current class as an event listener.
                .setActivity(Activity.customStatus("Helping
                to keep a friendly Discord server")) // Set
                the bot's custom status.
                .build();

        // Asynchronously get REST ping from Discord API
        and print it.
        jda.getRestPing().queue(ping -> System.out.
        println("Logged in with ping: " + ping));

        // Block the main thread until JDA is fully loaded.
        jda.awaitReady();
        // Print the number of guilds the bot is
        connected to.
        System.out.println("Guilds: " + jda.
        getGuildCache().size());
```

```
            // Print the Discord userID of the bot
            System.out.println("Bot's ID: " + jda.
            getSelfUser());
        } catch (InterruptedException e) {
            // Handle exceptions if the thread is interrupted
            during the awaitReady process.
            e.printStackTrace();
        }
    }

    @Override
    public void onMessageReceived(MessageReceivedEvent
    messageEvent){

        User senderDiscordID = messageEvent.getAuthor();
        MessageChannelUnion channel = messageEvent.
        getChannel();
        Message message = messageEvent.getMessage();

        // Check whether the message was sent in a guild
        / server
        if (messageEvent.isFromGuild()){

            String content = message.getContentDisplay();
            // Check if the message contains the banned word
            if (content.contains(bannedWord)){

                // Delete the message
                message.delete().queue();

                // Mention the user who sent the
                inappropriate message
                String authorMention = senderDiscordID.
                getAsMention();

                // Send a message mentioning the user and
                explaining why it was inappropriate
                channel.sendMessage(authorMention + " This
                comment was deemed inappropriate for this
                channel. " +
                        "If you believe this to be in error,
                        please contact one of the human server
                        moderators.").queue();
            }
        }
```

```
    }
  }
```

5.15.1　处理发送到 Discord 服务器的消息

让我们再次将注意力集中在 onMessageReceived()方法上，因为每次有消息发布到 Discord 服务器时，它都会被异步调用。正如你所看到的，如果发布到服务器的消息包含禁用词，就会删除消息，并在发布违规消息的同一频道中以@mention 消息警告发送者。

5.15.2　再次成功！运行第二个 Discord 机器人：ContentModeratorBotDumb.java

现在运行第二个 Java Discord 机器人。执行应用程序后，务必返回 Discord 服务器，并在任何包含违规词语的频道中键入一条消息。图 5-17 为机器人的运行情况。

图 5-17　该机器人对讨论“小狗”有严格规定，但讨论“小猫”却完全没有问题

5.16　小结

我们刚刚完成了用 Java 创建两个正常运行的 Discord 机器人的所有必要步骤。对于那些不熟悉创建 Discord 服务器过程的人，我们展示了如何设置服务器来管理社区。

正如你所见，与第 3 章中的 Slack 机器人相比，我们采用了截然不同的方法！创建的 Slack 机器人主要关注用户在工作场所的工作效率。而这两个 Discord 机器人则真正侧重于社区管理。在 OpenAI 应用程序接口的帮助下，已经为这些机器人实现人工智能做好了一切准备，而这些都将在最后两章中完成。

5.17 练习

在第 6 章中，我们将让“笨”机器人变得智能。不过，至少有一件事我们现在就可以做：与其使用命令行报告状态信息，不如让机器人拥有自己的频道，专门用于报告状态。这样，当机器人启动、关闭或有重要事情需要通知管理员时，所有信息都会被记录到一个中心位置。

第6章

为 Discord 机器人添加智能的第 1 部分：使用聊天端点进行问答

至此，我们已经具备了所有条件，能使我们在第 5 章中创建的两个 Discord 机器人都具有完整功能和人工智能。在本书的最后两章中，将按照所有必要的步骤使这两个机器人成为人工智能机器人。在本章中，将从名为 TechSupportBotDumb.java 的技术支持机器人开始。下面是要做的两个主要修改：

- 修改 ChatGPTClient.java 类，以便 Discord 机器人可以就我们提供给它的特定信息进行提问。更新后的类将称为 ChatGPTClientForQAandModeration.java。它将在本章中用于问答环节，但也会在本书的最后一章中使用。
- 修改 TechSupportBot.java 类(原名为 TechSupportBotDumb.java)，使其可以加载包含常见问题和答案的外部文本文件。然后，TechSupportBot.java 会将文本文件的内容提供给 ChatGPTClientForQAandModeration.java 类，后者负责创建提示，当然也会调用聊天端点。

6.1 使 TechSupportBot.java 更智能

代码清单 6-1 包含了所有常见的问题，这些问题是虚构的客户支持团队根据新推出的移动银行应用程序的用户的支持单据创建的。

代码清单 6-1 FAQ.txt

```
1. What is the Crooks Bank Mobile App?
The Crooks Bank Mobile App is a cutting-edge mobile banking app
that allows you to manage your finances, make transactions, and
access a wide range of banking services conveniently from your
mobile device.

2. How can I download the Crooks Bank Mobile App?
You can download the Crooks Bank Mobile App from the App
Store for iOS devices and Google Play for Android devices.
Simply search for the "Crooks Bank Mobile App" and click the
"Install" button.

3. Is the Crooks Bank Mobile App safe and secure?
Yes, the Crooks Bank Mobile App prioritizes your security. We use
state-of-the-art encryption and security protocols to protect
your data and transactions. Your information is safe with us.

4. What features does the Crooks Bank Mobile App offer?
The Crooks Bank Mobile App provides a variety of features,
including:
• Account Management: View account balances, transaction
history, and more.
• Transfer Funds: Easily transfer money between your
  accounts or to other bank accounts.
• Bill Payments: Pay bills and manage recurring payments.
• Deposit Checks: Snap photos of checks for remote deposit.
• ATM Locator: Find nearby ATMs and branches.
• Notifications: Receive alerts for account activity and
  important updates.

5. Can I link external accounts to the Crooks Bank Mobile App?
Yes, the Crooks Bank Mobile App supports linking external
accounts from other financial institutions. You can monitor and
manage your accounts from different banks in one place.
```

```
6. How can I reset my password if I forget it?
If you forget your password, simply click the "Forgot Password"
option on the login screen. You'll receive instructions on how
to reset your password.

7. What are the fees associated with the Crooks Bank
Mobile App?
The Crooks Bank Mobile App aims to be transparent with its
fees. You can find information on account fees, transaction
charges, and other costs in the "Fees" section within the app
or on our website.

8. Can I get customer support through the Crooks Bank
Mobile App?
Absolutely! We offer customer support through our in-app
messaging feature. You can also find our customer service
contact information on our website.

9. Is the Crooks Bank Mobile App available for business
accounts?
The Crooks Bank Mobile App primarily caters to personal banking
needs. However, we have plans to introduce business banking
services in the future.

10. How can I provide feedback or suggestions for the Crooks
Bank Mobile App?
We welcome your feedback! You can submit suggestions and
feedback through the "Contact Us" section in the app or on our
website.
```

中文注释

1. 什么是 Crooks Bank 移动应用程序？

Crooks Bank 移动应用程序是一款先进的移动银行应用程序，可让你通过移动设备方便地管理财务、进行交易和获取各种银行服务。

2. 如何下载 Crooks Bank 移动应用程序？

你可从iOS设备的App Store和Android设备的Google Play下载Crooks Bank 移动应用程序。只需要搜索 Crooks Bank Mobile App，然后单击 Install 按钮即可。

3. Crooks Bank 移动应用程序是否安全可靠？

当然，Crooks Bank 移动应用程序将安全放在首位。我们使用最先进的加密和安全协议来保护数据和交易，因此你的信息在我们这里是安全的。

4. Crooks Bank 移动应用程序提供哪些功能?

Crooks Bank 移动应用程序提供多种功能，包括

- 账户管理：查看账户余额、交易历史记录等。
- 转账：在你的账户之间轻松转账或转账至其他银行账户。
- 账单支付：支付账单并管理定期付款。
- 存入支票：拍摄支票照片，进行远程存款。
- ATM 定位器：查找附近的 ATM 和分行。
- 通知：接收账户活动和重要更新的提醒。

5. 可将外部账户链接到 Crooks Bank Mobile App 吗?

可以，Crooks Bank 移动应用程序支持链接其他金融机构的外部账户。你可在一个地方监控和管理来自不同银行的账户。

6. 如果忘记密码，如何重置?

如果忘记密码，只需要单击登录界面上的“忘记密码”选项。你将收到如何重设密码的说明。

7. Crooks Bank 移动应用程序的相关费用是多少?

Crooks Bank 移动应用程序的收费力求透明。你可以在应用程序内的“费用”部分或网站上找到有关账户费用、交易费用和其他费用的信息。

8. 我能通过 Crooks Bank 移动应用程序获得客户支持吗?

当然可以！我们通过应用程序内的消息功能提供客户支持。你也可以在我们的网站上找到客户服务联系信息。

9. Crooks Bank 移动应用程序是否适用于企业账户?

Crooks Bank 移动应用程序主要满足个人银行业务需求。不过，我们计划在未来推出企业银行服务。

10. 如何为 Crooks Bank 移动应用程序提供反馈或建议?

欢迎你提供反馈意见！你可通过应用程序中的“联系我们”或我们的网站提交建议和反馈。

正如你在代码清单 6-1 中的常见问题文本文件中所看到的，它并不涉及任何技巧，只是一个关于问题和答案的列表。现在，让我们看看新修改的 TechSupportBot.java 类。如代码清单 6-2 所示。

代码清单 6-2　TechSupportBot.java

```
import java.io.BufferedReader;
import java.io.FileReader;
import java.io.IOException;
import java.util.EnumSet;

import net.dv8tion.jda.api.JDA;
import net.dv8tion.jda.api.JDABuilder;
import net.dv8tion.jda.api.entities.Activity;
import net.dv8tion.jda.api.entities.User;
import net.dv8tion.jda.api.entities.channel.ChannelType;
import net.dv8tion.jda.api.entities.channel.unions.
MessageChannelUnion;
import net.dv8tion.jda.api.events.message.MessageReceivedEvent;
import net.dv8tion.jda.api.hooks.ListenerAdapter;
import net.dv8tion.jda.api.requests.GatewayIntent;

// This class extends a ListenerAdapter to handle message
events on Discord.
public class TechSupportBot extends ListenerAdapter {

    // The bot's Discord token for authentication.
    static String discordToken = "";
    // The name of the channel the bot should monitor and
    interact with.
    static String channelToWatch = "q-and-a";
    // Variable to store FAQ contents
    static String contentsFromFAQ = "";
    static String pathToFAQFile = "/Users/Desktop/FAQ.txt";
    // the system message
    static String systemMessage = "You are a virtual assistant
    that provides support for the Crooks Bank banking app. ".";
    // our ChatGPT client
    static ChatGPTClientForQAandModeration
    chatGPTClient = null;

    public static void main(String[] args) throws IOException {

        // Set of intents declaring which types of events the
        bot intends to listen to.
        EnumSet<GatewayIntent> intents = EnumSet.of(
```

```
            GatewayIntent.GUILD_MESSAGES, // For messages
            in guilds.
            GatewayIntent.DIRECT_MESSAGES, // For private
            direct messages.
            GatewayIntent.MESSAGE_CONTENT // To allow
            access to message content.
    );

    // Read the contents of an external text file into
    FAQContents variable
    contentsFromFAQ = readFileContents(pathToFAQFile);

    // create a new ChatGPTClientForQAandModeration
    chatGPTClient = new ChatGPTClientForQAandModeration(con
    tentsFromFAQ, systemMessage);

    // Initialize the bot with minimal configuration and
    the specified intents.
    try {
        JDA jda = JDABuilder.createLight(discordToken,
        intents)
                .addEventListeners(new TechSupportBot())
                // Adding the current class as an event
                listener.
                .setActivity(Activity.customStatus("Ready
                to answer questions")) // Set the bot's
                custom status.
                .build();

        // Asynchronously get REST ping from Discord API
        and print it.
        jda.getRestPing().queue(ping -> System.out.
        println("Logged in with ping: " + ping));

        // Block the main thread until JDA is fully loaded.
        jda.awaitReady();
        // Print the number of guilds the bot is
        connected to.
        System.out.println("Guilds: " + jda.
        getGuildCache().size());
        System.out.println("Self user: " + jda.
        getSelfUser());
    } catch (InterruptedException e) {
        // Handle exceptions if the thread is interrupted
        during the awaitReady process.
```

```
        e.printStackTrace();
    }
}

// This method handles incoming messages.
@Override
public void onMessageReceived(MessageReceivedEvent
messageEvent) {

    // The ID of the sender
    User senderDiscordID = messageEvent.getAuthor();
    // The Discord channel where the message was posted
    MessageChannelUnion channel = messageEvent.
    getChannel();
    net.dv8tion.jda.api.entities.Message message =
    messageEvent.getMessage();
    String reply = null;

    // Ignore messages sent by the bot to prevent selfresponses.
    if (senderDiscordID.equals(messageEvent.getJDA().
    getSelfUser())) {
        return;
} else if (messageEvent.getChannelType() ==
ChannelType.TEXT) {
    // Ignore messages not in the specified "q-and-a"
    channel.
    if (!channel.getName().equalsIgnoreCase(channelT
        oWatch)) {
            return;
        }
    }
    // Show "typing" status while the bot is working
    channel.sendTyping().queue();

        // this line takes the question from the Discord users
        and asks ChatGPT
        reply = chatGPTClient.sendMessageFromDiscordUser
        (message.getContentDisplay());
        channel.sendMessage(reply).queue();
    }

    // New method to read file contents
    private static String readFileContents(String filePath) {
        try (BufferedReader reader = new BufferedReader
        (new FileReader(filePath))) {
```

```
                StringBuilder content = new StringBuilder();
                String line;
                while ((line = reader.readLine()) != null) {
                    content.append(line).append("\n");
                }
                return content.toString();
            } catch (IOException e) {
                e.printStackTrace();
                return "Failed to read FAQ contents.";
            }
        }
    }
```

6.2 较之前一版本的技术支持机器人，需要注意的重要更改

让我们简要分析一下 TechSupportBot.java，并讨论一下所做的更改。下面的代码片段包含类定义的一部分。

```
static String contentsFromFAQ = "";
static String pathToFAQFile = "/Users/Desktop/FAQ.txt";
static String systemMessage = "You are a virtual assistant
that provides support for the Crooks Bank banking app. ";
static ChatGPTClientForQAandModeration
chatGPTClient = null;
```

正如你所看到的，我们定义了一些字符串，这些字符串提供了对存储 FAQ(常见问题)文件的文件路径位置的引用。我们还定义了一个字符串，用于包含文件本身的内容。

正如我们在本书前几章中学到的，通过在提示中向系统本身提供特定信息，可以确定对话的基调。因此，这里也有一个包含系统信息的字符串。最后，我们引用了一个类 ChatGPTClientForQAandModeration，它与我们之前在书中使用的其他 ChatGPTClient 类非常相似。

更新 onMessageReceived()方法

现在，当收到一条消息时，请务必注意以下代码：

```
net.dv8tion.jda.api.entities.Message message = messageEvent.
getMessage();
```

此处，由于我们已经创建并使用了 Message 类，以便在向聊天端点发送 HTTP 请求时封装和表示 JSON 对象，因此，我们需要给出 JDA 库使用的 Message 类的完整包和类名。

现在让我们进一步检查以下几行代码：

```
channel.sendTyping().queue();
reply = chatGPTClient.sendMessageFromDiscordUser(message.
getContentDisplay());
channel.sendMessage(reply).queue();
```

在这里，我们提供了良好的用户体验，并向用户展示了机器人正在“打字”，同时用户的问题正在被发送到 ChatGPT。若收到回复，我们就将其提供给用户。

6.3　分析 ChatGPTClientForQAandModeration.java

在代码清单 6-2 中，TechSupportBot.java 实例化了 ChatGPTClientForQAandModeration.java，它与我们之前用过的 ChatGPTClient 类非常相似。ChatGPTClientForQAandModeration.java 的完整源代码如代码清单 6-3 所示。

代码清单 6-3　ChatGPTClientForQAandModeration.java

```
import com.fasterxml.jackson.core.JsonProcessingException;
import com.fasterxml.jackson.databind.JsonNode;
import com.fasterxml.jackson.databind.ObjectMapper;

import java.io.BufferedReader;
import java.io.IOException;
import java.io.InputStreamReader;
import java.io.OutputStream;
```

```
import java.net.HttpURLConnection;
import java.net.URL;
import java.util.ArrayList;
import java.util.List;

public class ChatGPTClientForQAandModeration {

    //
    // OpenAI parameters that we already know how to use
    //
    String openAIKey = "";
    String endpoint = "https://api.openai.com/v1/chat/
    completions";
    String model = "gpt-4";
    float temperature = 1.0f;
    int max_tokens = 256;
    float top_p = 1.0f;
    int frequency_penalty = 0;
    int presence_penalty = 0;

    String systemMessage = null;
    String initialInstructionsToChatGPT = null;
    //
    // The constructor needs to be passed the contents from the
    FAQ.txt file
    // and the system message
    //
    public ChatGPTClientForQAandModeration(String
    systemMessage, String initialInstructionsToChatGPT) {
        this.systemMessage = systemMessage;
        this.initialInstructionsToChatGPT =
        initialInstructionsToChatGPT;
    }

    public String sendMessageFromDiscordUser(String
    discordMessageText) {

        String answerFromChatGPT = "";

        List<Message> messages = new ArrayList<>();
        messages.add(new Message("system", systemMessage));
        messages.add(new Message("user",
        initialInstructionsToChatGPT));
        messages.add(new Message("user", discordMessageText));
```

```
    String jsonInput = null;
    try {
        ObjectMapper mapper = new ObjectMapper();

        Chat chat = Chat.builder()
            .model(model)
            .messages(messages)
            .temperature(temperature)
            .maxTokens(max_tokens)
            .topP(top_p)
            .frequencyPenalty(frequency_penalty)
            .presencePenalty(presence_penalty)
            .build();

    jsonInput = mapper.writeValueAsString(chat);
    System.out.println(jsonInput);
} catch (JsonProcessingException e) {
    e.printStackTrace();
}

try {
    URL url = new URL(endpoint);
    HttpURLConnection connection = (HttpURLConnection)
    url.openConnection();
    connection.setRequestMethod("POST");
    connection.setRequestProperty("Content-Type",
    "application/json");
    connection.setRequestProperty("Authorization",
    "Bearer " + openAIKey);
    connection.setDoOutput(true);

    OutputStream outputStream = connection.
    getOutputStream();
    outputStream.write(jsonInput.getBytes());
    outputStream.flush();
    outputStream.close();

    int responseCode = connection.getResponseCode();
    if (responseCode == HttpURLConnection.HTTP_OK) {
        BufferedReader reader = new BufferedReader(new
        InputStreamReader(connection.
        getInputStream()));
        StringBuilder response = new StringBuilder();
            String line;
            while ((line = reader.readLine()) != null) {
```

```
                response.append(line);
            }
            reader.close();

            // Print the response
            answerFromChatGPT =
            extractAnswerFromJSON(response.toString());
            System.out.println(answerFromChatGPT);
        } else {
            System.out.println("Error: " + responseCode);
        }
        connection.disconnect();
    } catch (IOException e) {
        e.printStackTrace();
    }
    return answerFromChatGPT;
}

//
// We are only interested in the "message.content" in the
    JSON response
// So here's the easy way to extract that
//
public String extractAnswerFromJSON(String jsonResponse) {
    String chatGPTAnswer = "";

    try {
        // Create an ObjectMapper instance
        ObjectMapper objectMapper = new ObjectMapper();

        // Parse the JSON string
        JsonNode rootNode = objectMapper.
        readTree(jsonResponse);

        // Extract the "content" parameter
        JsonNode contentNode = rootNode.
        at("/choices/0/
        message/content");
        chatGPTAnswer = contentNode.asText();

        System.out.println("Content: " + chatGPTAnswer);

    } catch (Exception e) {
        e.printStackTrace();
    }
```

```
        return chatGPTAnswer;
    }
}
```

其中最重要的一点是，在构造函数中，我们要将 FAQ 内容的完整字符串以及要提供给系统本身的信息发送出去。

```
public ChatGPTClientForQAandModeration(String
knowledgeBaseFileContents, String systemMessage) {
        this.knowledgeBaseFileContents =
        knowledgeBaseFileContents;
        this.systemMessage = systemMessage;
    }
```

这样，构建了 ChatGPTClientForQAandModeration.java 类后，我们就可以重复使用已构建的对象，以便向用户询问每个问题。

每次有问题发布到 Discord 频道上时，TechSupportBot.java 就会调用 ChatGPTClientForQAandModeration.java 中的 sendMessageFromDiscordUser() 方法。在该方法中，我们将执行调用聊天端点需要的所有常规操作。

6.3.1　使用 JSONPath 快速提取 JSON 文件中的内容

有时，为了提取 JSON 文件中的单个节点，编写必要的代码可能会有点乱。因此，我使用 JSONPath 编写了一个简单的方法，以便在聊天端点返回的 JSON 文件中提取 “message.content” 节点。而这一切只需要一行代码即可完成：

```
JsonNode contentNode = rootNode.at("/choices/0/message/
content");
```

在我看来，JsonPath 对于从 JSON 文件中提取单个节点或节点组非常有效，但表达式本身和必要的语法就像正则表达式一样复杂。因此，如果你以前从未用过 JSONPath，或者在如何编写表达式方面需要帮助，那么请使用 ChatGPT 结对编程！

6.3.2　运行智能问答机器人：TechSupportBot.java

现在让我们运行新的改进型问答 Java Discord 机器人 TechSupport

Bot.java。执行应用程序后，务必返回 Discord 服务器并开始提问。图 6-1 是机器人的运行情况。

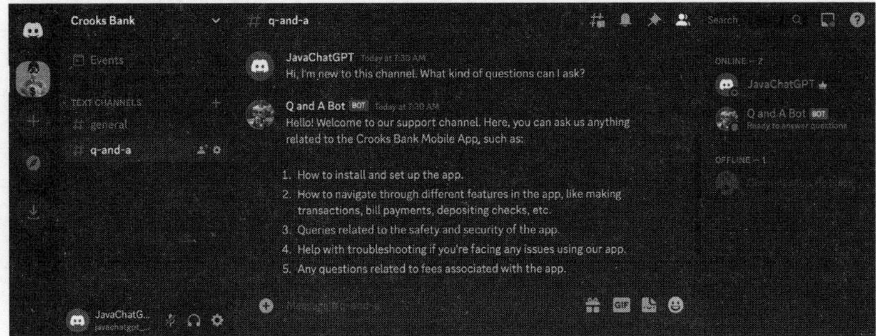

图 6-1 与智能 Discord 问答机器人 TechSupportBot.java 进行讨论

代码清单 6-4 是我们与 Discord 机器人之间的对话。

代码清单 6-4 与智能 Discord 问答机器人的第一次对话

```
Me: Hi, I'm new to this channel. What kind of questions
can I ask?

Q and A Bot: Hello! Welcome to our support channel. Here, you
can ask us anything related to the Crooks Bank Mobile App,
such as:

1. How to install and set up the app.
2. How to navigate through different features in the app, like
making transactions, bill payments, depositing checks, etc.
3. Queries related to the safety and security of the app.
4. Help with troubleshooting if you're facing any issues using
our app.
5. Any questions related to fees associated with the app.
Don't hesitate to ask us anything else related to our services!
We're here to assist you.

Me: I have an old Blackberry phone, is the app supported on
my device?

Q and A Bot: I'm sorry for the inconvenience, but at the
moment, the Crooks Bank Mobile App is currently only available
for iOS and Android devices. It is not supported on Blackberry
or any other operating systems. You may still manage your bank
```

```
accounts by accessing our website through your phone's browser.

Me: So, if I use an iPhone, how do I setup the app?

Q and A Bot: Sure, here's how you can set up the Crooks Bank
Mobile App on your iPhone:

1. Open the App Store on your iPhone.
2. In the search bar at the top, type in "Crooks Bank Mobile
App" and tap "Search".
3. Once you can see the app in the search results, simply tap
on the "Get" button to download it.
4. After the app downloads, tap "Open".
5. You'll have to agree to our Terms and Conditions upon the
first launch.
6. If you're an existing Crooks Bank member, you can log in
using your online banking user ID and password. If you're a new
member, you'll need to register for an account.
7. Once you're logged in, you can set up Touch ID or Face ID
for quicker, secure login in the future.
8. Now you can start using your Crooks Bank Mobile App to
manage your finances, pay bills, make transactions, and more!

Please make sure you update the app regularly to access new
features and improvements. If you have any challenges during
setup, please reach out to us through the app's support feature
or call our customer service.
```

中文注释

我：你好，我是这个频道的新用户。请问我可以提出什么问题？

问答机器人：你好！欢迎来到我们的支持频道。在这里，你可以向我们提出任何与 Crooks Bank 移动应用程序有关的问题，例如：

1. 如何安装和设置应用程序。

2. 如何浏览应用程序中的不同功能，如进行交易、支付账单、存入支票等。

3. 与应用程序的安全有关的问题。

4. 如果你在使用应用程序时遇到任何问题，将帮助你排除故障。

5. 与应用程序相关费用有关的任何问题。

如果你对我们的服务有任何疑问，不要犹豫！我们将竭诚为你服务。

我：我有一部旧的黑莓手机，我的设备支持该应用程序吗？

问答机器人：很抱歉给你带来不便，但目前克鲁克斯银行移动应用程序仅适用于 iOS 和 Android 设备。它不支持黑莓或任何其他操作系统。你仍然可以通过手机浏览器访问我们的网站来管理你的银行账户。

我：如果使用的是 iPhone，我该如何设置应用程序？

问答机器人：在 iPhone 上设置 Crooks Bank 移动应用程序的方法如下。

1. 打开 iPhone 上的 App Store。
2. 在顶部的搜索栏中输入 Crooks Bank Mobile App，然后单击 Search。
3. 在搜索结果中看到该应用程序后，只需要单击 Get 按钮即可下载。
4. 应用程序下载完成后，单击 Open。
5. 首次启动时，你必须同意我们的条款和条件。
6. 如果你是 Crooks Bank 的现有会员，你可以使用网上银行用户名和密码登录。如果你是新会员，则需要注册账户。
7. 登录后，你可以设置 Touch ID 或 Face ID，以便将来更快速、安全地登录。
8. 现在你可以开始使用 Crooks Bank 移动应用程序管理你的财务、支付账单、进行交易等！

请确保定期更新应用程序，以使用新功能和改进功能。如果你在设置过程中遇到任何问题，请通过应用程序的支持功能联系我们或致电我们的客户服务。

6.4 我们取得了巨大成就，但有一个小缺陷

好吧，如果你退一步审视我们迄今为止所取得的成就，你就会意识到我们取得了不小的成果：

- 一个由几个类组成的系统，允许用户输入问题，并获得有关如何使用移动应用程序的答案。
- 通过一个简单的文本文件，我们可以教会机器人如何回答用户的问题。公司里的任何人都可以编辑这个文件，并将其作为知识库，帮助机器人不断提高智能。这真了不起！

- 该系统允许客户使用自然语言输入问题，然后机器人会向他们提供智能回答。你猜怎么着？客户不喜欢阅读常见问题，尤其是很长的问题。但是，使用该系统后，他们就不需要这样做了！他们只需要提出与自己相关的问题即可。

因此，完成所有这些工作之后，有一个严重的缺陷是我们不能忽视的。在代码清单 6-4 中，机器人对用户说：

```
Once you're logged in, you can set up Touch ID or Face ID for
quicker, secure login in the future.
```

中文注释

“登录后，你可以设置 Touch ID 或 Face ID，以便将来更快、更安全地登录。”

不，不，不！坏机器人！如果你没有阅读完常见问题文件，请允许我解释一下这里出了什么问题：

(1) 代码清单 6-1 中的 FAQ.txt 文件明确指出 Crooks Bank 移动应用程序是一款移动应用程序。目前，Touch ID 只能在苹果台式机和笔记本电脑上使用。所以这根本说不通。

(2) 在 FAQ.txt 文件中，根本没有提到 Face ID 可用于手机应用的身份验证。

因此，遗憾的是，ChatGPT 捏造了一些事实，并假定某些功能存在，但其实际上并不存在。因此，就真正的技术支持而言，这是绝对不能接受的。

那么，该如何解决这个问题呢？在阅读下一部分之前，请先想一想如何解决这个问题。

6.5　将系统信息更新为 ChatGPT，再试一次

你想出解决方案了吗？解决这个问题的方法之一就是更新系统消息。请参阅代码清单 6-2 中的以下一行代码：

```
static String systemMessage = "You are a virtual assistant that
```

```
provides support for the Crooks Bank banking app. ";
```

中文注释

static String systemMessage = “你是为 Crooks Bank 银行应用程序提供支持的虚拟助手。”；

请记住，系统消息向 ChatGPT 表明了它在聊天对话中的角色。现在，它被告知担任一名提供帮助的助手，因此，它擅自猜测了移动应用程序中存在的某些功能。现在，让我们确保 ChatGPT 不会提供任何假设。更新系统消息，如下所示：

```
static String systemMessage = "You are a virtual assistant
that provides support for the Crooks Bank banking app. You
will be provided Information about the application, and users
will ask you questions about the application. Your role is to
answer questions explicitly from the training information. If
you were not given an answer to a question, then provide an
apologetic message to the user explaining that you do not know
the answer.";
```

中文注释

static String systemMessage=“你是为 Crooks Bank 银行应用程序提供支持的虚拟助手。你将获得有关应用程序的信息，用户会向你提出有关应用程序的问题。你的职责是明确回答培训信息中的问题。如果你不知道某个问题的答案，那么请向用户道歉和解释，表明你不知道答案”；

就是这样。FAQ.txt 文件与之前的完全相同，我们需要做的就是更新系统消息，让 ChatGPT 知道自己在对话中的角色。代码清单 6-5 是我们和 Discord 机器人之间的新对话。

代码清单 6-5　与智能问答 Discord 机器人的第二次对话的效果更好

```
Me: How are you today?

Q and A Bot: As a virtual assistant, I don't have feelings or
emotions, but I'm here and ready to assist you. How may I help
you with the Crooks Bank Mobile App today?
```

Me: I have an old Blackberry phone, is the app supported on my device?

Q and A Bot: I'm sorry, but from the information provided, the Crooks Bank Mobile App can only be downloaded from the App Store for iOS devices and Google Play for Android devices. BlackBerry devices are not mentioned, therefore I cannot confirm that it would be compatible.

Me: So, if I use an iPhone, how do I setup the app?

Q and A Bot: To set up the Crooks Bank Mobile App on your iPhone, you can download it from the App Store. Simply search for "Crooks Bank Mobile App" and click the "Install" button. After the installation is finished, open the app and follow the prompts to set up your account. If you already have an account, use your login information to access your account through the app.

中文注释

我： 你今天好吗？

问答机器人： 作为一个虚拟助理，我没有感情或情绪，但我在这里随时准备为你提供帮助。今天我可以通过 Crooks Bank 移动应用程序为你提供哪些帮助？

我： 我有一部旧的黑莓手机，我的设备支持该应用程序吗？

问答机器人： 很抱歉，根据所提供的信息，克鲁克银行移动应用程序只能在 iOS 设备的 App Store 和 Android 设备的 Google Play 下载。没有提到黑莓设备，因此我无法确认它是否兼容。

我： 那么，如果我使用 iPhone，该如何设置应用程序？

问答机器人： 要在 iPhone 上设置 Crooks Bank 移动应用程序，可以从 App Store 下载。只需要搜索 Crooks Bank Mobile App，然后单击“安装”按钮。安装完成后，打开应用程序并按提示设置账户。如果你已经有一个账户，请使用你的登录信息通过应用程序访问账户。

现在好多了！我们的技术支持机器人“按部就班”，不允许虚构或假设任何事情。

6.6 小结

在本章中，我们已经完成了很多工作！我们有了一个功能完备的Discord 机器人，没有任何人工智能、NLP 或机器学习经验的人都可以使用一个简单的文本文件对它进行训练。我们了解到，“得力助手”有时可能变得过于得力。我们也重申了系统消息的概念和价值，这是提示工程的重要组成部分。

既然我们已经让问答 Discord 机器人变得智能了，现在看看如何让内容审核机器人也变得智能！

第7章

为 Discord 机器人添加智能的第 2 部分：使用聊天和审核端点进行审核

在本章中，我们将采取必要步骤，使内容审核 Discord 机器人具有人工智能。让我们概述一下要做的更改。

- 创建一个新类 ModerationClient.java 以调用审核端点。当任何文本内容符合以下任何类别时，审核(Moderation)端点都能让我们意识到：
    - 仇恨
    - 仇恨/威胁
    - 骚扰
    - 骚扰/威胁
    - 自我伤害
    - 自残：意图
    - 自残/指令
    - 性行为
    - 性/未成年人
    - 暴力

- 暴力/图形
- 重复使用上一章中的 ChatGPTClientForQAandModeration.java。在第 6 章中，它被用来调用聊天端点，以便用户进行问答。在本章中，它将再次用于调用聊天端点，但这次用于审核。
- 修改 ContentModeratorBot.java 类(原名 ContentModeratorBot Dumb.java)，使其可以同时调用 ModerationClient.java 和 ChatGPT ClientFor QAandModeration.java。如果这两个类都显示在 Discord 频道中键入的内容令人不适，就会从该 Discord 频道中删除信息。记住，机器人会监视 Discord 服务器所有频道中的所有内容！

注意：

现在，说到这里，你可能会问自己，既然管理端点已经知道如何标记任何不适内容，那为什么我们还需要使用聊天端点呢？问得好。

是的，"审核端点"可以让我们了解有害内容，但不能通知我们任何其他类型的不受欢迎的内容，例如，当不法分子试图引诱我们的用户上当受骗时。请记住，这是一个银行应用程序的 Discord 服务器，因此骗子肯定会盯上这个 Discord 服务器的所有成员，因为这里是银行用户集中的地方！

因此，我们将使用 ModerationClient.java 来调用审核端点，以了解 Discord 服务器中是否有任何有害内容，我们还将重复使用第 6 章中的 ChatGPTClientForQAandModeration.java，以调用聊天端点，从而了解 Discord 服务器中发布的其他任何不良内容，如诈骗企图。

7.1 审核端点

审核端点允许开发人员提交文本字符串，并随后了解其是否具有暴力、仇恨、威胁或任何形式的骚扰内容。

7.1.1 创建请求

表 7-1 列出了调用审核端点需要的所有 HTTP 参数。

表 7-1　审核端点的 HTTP 参数

HTTP 参数	描述
Endpoint URL	https://api.openai.com/v1/moderations
Method	POST
Header	Authorization：Bearer **$OPENAI_API_KEY**
Content-Type	application/json

7.1.2　创建审核(JSON)

表 7-2 描述了审核端点请求体所需的 JSON 对象格式。该服务的使用非常简单，只需要一个参数即可正确调用该服务。

表 7-2　审核端点的请求正文

字段	类型	是否必需?	说明
Input	字符串或数组	必需	需要分类的文本
Model	字符串 默认值： text-moderation-latest	可选	实际上有两种内容审核模式可供使用：text-moderation-stable 和 text-moderation-latest。 默认设置为 text-moderation-latest。它会随着时间的推移自动升级，确保你始终使用最准确的模型。 如果使用 text-moderation-stable，则会在模型更新前提前通知你。 text-moderation-stable 的准确度往往略低于 text-moderation-latest

7.1.3　处理 JSON 响应

成功调用审核端点后，服务将提供一个 JSON 响应，其结构如表 7-3 所示。

表 7-3 审核 JSON 对象的结构

字段	类型	说明
Id	String	审核请求的唯一标识符
Model	String	用于执行审核请求的模型
Results	Array	审核对象列表
↳flagged	Boolean	标记内容是否违反 OpenAI 使用政策
↳categories	Array	类别及其是否被标记的列表
↳↳hate	Boolean	这表明所给出的文本是否表达、煽动或宣扬有关种族、性别、宗教、民族、国籍、残疾状况、性取向或种姓的仇恨
↳↳hate/threatening	Boolean	这表示所给文本是否包含仇恨内容，是否还基于上述偏见威胁要对目标群体使用暴力或造成严重伤害
↳↳harassment	Boolean	这表示所给文本是否包含对任何目标群体表达、煽动或宣扬骚扰性语言的内容
↳↳harassment/threatening	Boolean	这表示所给文本是否包含对任何目标群体威胁使用暴力或严重伤害的骚扰内容
↳↳self-harm	Boolean	这表示所提供的文本是否包含提倡、鼓励或描述自我伤害行为(如自杀、切腹和饮食失调)的内容
↳↳self-harm/intent	Boolean	这表示所给文本是否包含说话人表示他们正在或打算进行自杀、切腹和饮食失调等自我伤害行为的内容
↳↳self-harm/instructions	Boolean	这表示所给文本是否包含鼓励实施自杀、切腹和饮食失调等自我伤害行为的内容。 这包括指示或建议如何实施此类行为的内容
↳↳sexual	Boolean	这表示所给文字是否包含旨在引起性兴奋的内容，如对性活动的描述 这包括推广性服务的内容；但不包括性教育和健康等主题

(续表)

字段	类型	说明
↳↳sexual/minors	Boolean	这表示所给文本是否包含涉及 18 岁以下个人的内容
↳↳violence	Boolean	表示所给文本是否包含描述死亡、暴力或身体伤害的内容
↳↳violence/graphic	Boolean	表示是否包含描写死亡、暴力或身体伤害的图形细节
↳category_scores	Array	类别列表以及模型给出的分数
↳↳hate	Number	“仇恨”类别得分
↳↳hate/threatening	Number	“仇恨/威胁”类别得分
↳↳harassment	Number	“骚扰”类别得分
↳↳harassment/threatening	Number	“骚扰/威胁”类别得分
↳↳self-harm	Number	“自我伤害”类别得分
↳↳self-harm/intent	Number	“自我伤害/意图”类别得分
↳↳self-harm/instructions	Number	“自我伤害/指示”类别得分
↳↳sexual	Number	“性”类别得分
↳↳violence	Number	“暴力”类别得分
↳↳violence/graphic	Number	“暴力/图形”类别得分

注：表中的↳指示数组元素。

7.1.4　审核(JSON)

代码清单 7-1 是调用审核端点后的 JSON 响应示例。表 7-3 看起来有点复杂，但正如你所看到的，如果任何类别被标记为 true，那么 results.flagged 节点就会被标记为 true。

代码清单 7-1 是有关审核 JSON 对象的实际示例。

代码清单 7-1 审核 JSON 对象

```
{
    "id": "modr-XXXXX",
    "model": "text-moderation-005",
    "results": [
        {
        "flagged": true,
        "categories": {
        "sexual": false,
        "hate": false,
        "harassment": false,
        "self-harm": false,
        "sexual/minors": false,
        "hate/threatening": false,
        "violence/graphic": false,
        "self-harm/intent": false,
        "self-harm/instructions": false,
        "harassment/threatening": true,
        "violence": true,
        },
        "category_scores": {
        "sexual": 1.2282071e-06,
        "hate": 0.010696256,
        "harassment": 0.29842457,
        "self-harm": 1.5236925e-08,
        "sexual/minors": 5.7246268e-08,
        "hate/threatening": 0.0060676364,
        "violence/graphic": 4.435014e-06,
        "self-harm/intent": 8.098441e-10,
        "self-harm/instructions": 2.8498655e-11,
        "harassment/threatening": 0.63055265,
        "violence": 0.99011886,
        }
        }
    ]
}
```

7.2　为审核端点创建客户端：ModerationClient.java

代码清单 7-2 是调用审核端点的客户端。请先看一看，然后我们将讨论重要部分。

代码清单 7-2　ModerationClient.java

```
import com.fasterxml.jackson.core.JsonProcessingException;
import com.fasterxml.jackson.databind.JsonNode;
import com.fasterxml.jackson.databind.ObjectMapper;

import java.io.BufferedReader;
import java.io.IOException;
import java.io.InputStreamReader;
import java.io.OutputStream;
import java.net.HttpURLConnection;
import java.net.URL;
import java.util.ArrayList;

public class ModerationClient {
   // OpenAI parameters that we already know how to use
   String openAIKey = "";
   String endpoint = "https://api.openai.com/v1/moderations";
   String model = "text-moderation-latest";

   // The constructor
   public ModerationClient() {
   }

   public ModerationResponse checkForObjectionalContent(String
   discordMessageText) {

      ModerationResponse moderationResponse = null;

      String jsonInput = null;
      try {
         ObjectMapper mapper = new ObjectMapper();

         ModRequest modRequest = new
         ModRequest(discordMessageText, model);

         jsonInput = mapper.writeValueAsString(modRequest);
         System.out.println(jsonInput);
```

```
    } catch (JsonProcessingException e) {
        e.printStackTrace();
    }

    try {
        URL url = new URL(endpoint);
        HttpURLConnection connection = (HttpURLConnection)
        url.openConnection();
        connection.setRequestMethod("POST");
        connection.setRequestProperty("Content-Type",
        "application/json");
        connection.setRequestProperty("Authorization",
        "Bearer " + openAIKey);
        connection.setDoOutput(true);

        OutputStream outputStream = connection.
        getOutputStream();
        outputStream.write(jsonInput.getBytes());
        outputStream.flush();
        outputStream.close();

        int responseCode = connection.getResponseCode();
        if (responseCode == HttpURLConnection.HTTP_OK) {
            BufferedReader reader = new BufferedReader
            (new InputStreamReader(connection.
            getInputStream()));
            StringBuilder response = new StringBuilder();
            String line;
            while ((line = reader.readLine()) != null) {
                response.append(line);
            }
            reader.close();

            // Print the response
            //System.out.println(response.toString());
            // Extract the answer from JSON
            moderationResponse = getModerationResponsefrom
            JSON(response.toString());
            String answerFromChatGPT = moderationResponse.
            toString();
            System.out.println(answerFromChatGPT);
        } else {
            System.out.println("Error: " + responseCode);
        }
```

```
            connection.disconnect();
        } catch (IOException e) {
            e.printStackTrace();
        }
        return moderationResponse;
    }

    public ModerationResponse getModerationResponsefromJSON
    (String jsonResponse) {
        ModerationResponse response = new ModerationResponse();
        ObjectMapper mapper = new ObjectMapper();
        try {
            JsonNode rootNode = mapper.readTree(jsonResponse);
            JsonNode resultsNode = rootNode.path("results");
            if (!resultsNode.isMissingNode() && resultsNode.
            isArray() && resultsNode.size() > 0) {
                JsonNode resultNode = resultsNode.get(0);
                response.isFlagged = resultNode.
                path("flagged").asBoolean(false);
                JsonNode categoriesNode = resultNode.
                path("categories");
                if (!categoriesNode.isMissingNode()) {
                    categoriesNode.fields().
                    forEachRemaining(entry -> {
                        if (entry.getValue().
                        asBoolean(false)) {
                            response.offendingCategories.
                            add(entry.getKey());
                        }
                    });
                }
            }
    } catch (JsonProcessingException e) {
        e.printStackTrace();
    }
    return response;
}

class ModerationResponse {
    boolean isFlagged = false;
    ArrayList<String> offendingCategories = new
    ArrayList<>();

    @Override
```

```
        public String toString() {
            return "ModerationResponse{" +
                    "isFlagged=" + isFlagged +
                    ", offendingCategories=" +
                    offendingCategories +
                    '}';
        }
    }
}
```

在本书的前几章中，我们为 OpenAI API 的其他端点创建了客户端，因此上面的类看起来应该很熟悉。不过，在类的末尾，有一个名为 ModerationResponse 的内部类。

```
class ModerationResponse {
    boolean isFlagged = false;
    ArrayList<String> offendingCategories = new
    ArrayList<>();
```

该类封装了从审核端点返回的审核 JSON 对象中的有价值信息。也就是说，如果要评估的原始 Discord 消息违反了内容规则，会用布尔值 isFlagged 来通知我们。如果 isFlagged 为 true，那么 offendingCategories 中就会填入其内容被标记的类别。

因此，getModerationResponsefromJSON()方法的作用和其名字一样。我们传递由审核端点返回的审核 JSON 对象，然后得到一个完全实例化的 ModerationResponse 对象。

7.3　让 ContentModeratorBot.java 更智能

现在有了 ModerationClient.java 来调用审核端点，让我们看看更新后的 ContentModeratorBot.java(原名为 ContentModeratorBotDumb.java)，它将使用 ModerationClient.java 检查有害内容，并使用 ChatGPTClientForQAandModeration.java(上一章未修改)检查潜在的欺诈行为。

代码清单 7-3 是智能 Discord 审核机器人的完整源代码 ContentModeratorBot.java。

代码清单 7-3　ContentModeratorBot.java

```
import java.io.IOException;
import java.util.EnumSet;

import net.dv8tion.jda.api.JDA;
import net.dv8tion.jda.api.JDABuilder;
import net.dv8tion.jda.api.entities.Activity;
import net.dv8tion.jda.api.entities.User;
import net.dv8tion.jda.api.entities.channel.unions.
MessageChannelUnion;
import net.dv8tion.jda.api.events.message.MessageReceivedEvent;
import net.dv8tion.jda.api.hooks.ListenerAdapter;
import net.dv8tion.jda.api.requests.GatewayIntent;

// This class extends a ListenerAdapter to handle message
events on Discord.
public class ContentModeratorBot extends ListenerAdapter {

    // The bot's Discord token for authentication.
    static String discordToken = "";

    // the system message
    // This is a Java 13+ Multiline String notation. At the end
    of the day, it's still a String
    static String systemMessage = """

        You are the automated moderator assistant for a
        Discord server.
        Review each message for the following rule violations:
        1. Sensitive information
        2. Abuse
        3. Inappropriate comments
        4. Spam, for example; a message in all capital
           letters, the same phrase or word being repeated
           over and over, more than 3 exclamation marks or
           question marks.
        5. Advertisement
        6. External links
        7. Political messages or debate
        8. Religious messages or debate

        If any of these violations are detected, respond with
        "FLAG" (in uppercase without quotation marks). If the
        message adheres to the rules, respond with "SAFE" (in
```

```
    uppercase without quotation marks).
    """;

static String instructionsToChatGPT = "Analyze the
following message for rule violations:";

// this is our Chat Endpoint client
static ChatGPTClientForQAandModeration
chatGPTClient = null;
// this is our Moderations Endpoint client
static ModerationClient moderationClient = null;

public static void main(String[] args) throws IOException {

    // Set of intents declaring which types of events the
    bot intends to listen to.
    EnumSet<GatewayIntent> intents = EnumSet.of(
            GatewayIntent.GUILD_MEMBERS, // to get access
            to the members of the Discord server
            GatewayIntent.GUILD_MODERATION, // to ban and
            unban members
            GatewayIntent.GUILD_MESSAGES, // For messages
            in guilds
            GatewayIntent.MESSAGE_CONTENT // To allow
            access to message content
    );

    // create a new ChatGPTClientForQAandModeration
    chatGPTClient = new ChatGPTClientForQAandModeration
    (systemMessage, instructionsToChatGPT);
    // create a new ModerationClient
    moderationClient = new ModerationClient();

    // Initialize the bot with minimal configuration and
    the specified intents.
    try {
        JDA jda = JDABuilder.createLight(discordToken,
        intents)
                .addEventListeners(new
                ContentModeratorBot()) // Adding the
                current class as an event listener.
                .setActivity(Activity.customStatus("Helping
                to keep a friendly Discord server")) // Set
                the bot's custom status.
                .build();
```

```
            // Asynchronously get REST ping from Discord API
            and print it.
            jda.getRestPing().queue(ping -> System.out.
            println("Logged in with ping: " + ping));

            // Block the main thread until JDA is fully loaded.
            jda.awaitReady();

            // Print the number of guilds the bot is
            connected to.
            System.out.println("Guilds: " + jda.
            getGuildCache().size());
            // Print the Discord userID of the bot
            System.out.println("Bot's ID: " + jda.
            getSelfUser());
        } catch (InterruptedException e) {
            // Handle exceptions if the thread is interrupted
            during the awaitReady process.
            e.printStackTrace();
        }
    }

    @Override
    public void onMessageReceived(MessageReceivedEvent
    messageEvent){

        String chatGPTResponse = "";
        ModerationClient.ModerationResponse
        moderationResponse = null;
        User senderDiscordID = messageEvent.getAuthor();

        // The Discord channel where the message was posted
        MessageChannelUnion channel = messageEvent.
        getChannel();
        net.dv8tion.jda.api.entities.Message message =
        messageEvent.getMessage();

        // Ignore messages sent by the bot to prevent selfresponses.
        if (senderDiscordID.equals(messageEvent.getJDA().
        getSelfUser())) {
            return;
        }

        // this line takes the message from the Discord user
```

```
        and invokes the Moderation Endpoint
        moderationResponse = moderationClient.checkForObjection
        alContent(message.getContentDisplay());

        // this line takes the message from the Discord user
        and invokes the Chat Endpoint
        chatGPTResponse = chatGPTClient.sendMessageFromDiscordU
        ser(message.getContentDisplay());

        // Check whether the message was sent in a guild
        / server
        if (messageEvent.isFromGuild()){

            // Check both the Chat Endpoint and Moderation
            Endpoint to see if the message is flagged

            if (chatGPTResponse.equals("FLAG") ||
            moderationResponse.isFlagged ){

                // Delete the message
                message.delete().queue();

                // Mention the user who sent the
                inappropriate message
                String authorMention = senderDiscordID.
                getAsMention();

                // Send a message mentioning the user and
                explaining why it was inappropriate
                channel.sendMessage(authorMention + " This
                comment was deemed inappropriate for this
                channel. " +
                        "If you believe this to be in error,
                        please contact one of the human server
                        moderators.").queue();
            }
        }
    }
}
```

7.4　与上一版内容审核机器人相比，应注意的重要更改

让我们简要看看代码清单 7-3 中的 ContentModeratorBot.java 并讨论一下所做的更改。以下代码片段包含类定义的一部分。

```
static String systemMessage = """
      You are the automated moderator assistant for a
      Discord server.
      Review each message for the following rule violations:
      1. Sensitive information
      2. Abuse
      3. Inappropriate comments
      4. Spam, for example; a message in all capital
         letters, the same phrase or word being repeated
         over and over, more than 3 exclamation marks or
         question marks.
      5. Advertisement
      6. External links
      7. Political messages or debate
      8. Religious messages or debate

      If any of these violations are detected, respond with
      "FLAG" (in uppercase without quotation marks). If the
      message adheres to the rules, respond with "SAFE" (in
      uppercase without quotation marks).
      """;

   static String instructionsToChatGPT = "Analyze the
   following message for rule violations:";
   // this is our Chat Endpoint client
   static ChatGPTClientForQAandModeration
   chatGPTClient = null;
   // this is our Moderation Endpoint client
   static ModerationClient moderationClient = null;
```

中文注释

static String systemMessage = """

你是 Discord 服务器的自动审核助理。

```
检查每条消息是否违反以下规则:
1. 敏感信息
2. 滥用
3. 不当评论
4. 垃圾邮件，例如：全大写字母的信息、重复使用同一短语或单词、超过 3 个感叹号或问号。
5. 广告
6. 外部链接
7. 政治信息或辩论
8. 宗教信息或辩论
如果发现任何上述违规行为，请回复 FLAG(大写，不含引号)。如果信息符合规定，则回复 SAFE(大写，不含引号)。
"""";
static String instructionsToChatGPT = "分析以下信息是否违反规则:";
// this is our Chat Endpoint client
static ChatGPTClientForQAandModeration
chatGPTClient = null;
// this is our Moderation Endpoint client
static ModerationClient moderationClient = null;
```

如果你使用的是 Java 13+，则可以使用“三引号”符号定义整个文本块。我们就是这样定义 ChatGPTClientForQAandModeration 类将使用的系统消息的。

更新 onMessageReceived( )方法

Discord 服务器的任何频道收到消息后，都会调用 onMessageReceived()方法。以下是需要注意的最重要更改：

```
moderationResponse = moderationClient.checkForObjection
alContent(message.getContentDisplay());

chatGPTResponse = chatGPTClient.sendMessageFromDiscord
User(message.getContentDisplay());
```

```
// Check whether the message was sent in a guild
/ server
if (messageEvent.isFromGuild()){

    // Check both the Chat Endpoint and Moderation
    Endpoint to see if the message is flagged

    if (chatGPTResponse.equals("FLAG") ||
    moderationResponse.isFlagged ){

        // Delete the message
        message.delete().queue();

        // Mention the user who sent the
        inappropriate message
        String authorMention = senderDiscordID.
        getAsMention();

        // Send a message mentioning the user and
        explaining why it was inappropriate
        channel.sendMessage(authorMention + " This
        comment was deemed inappropriate for this
        channel. " +
                "If you believe this to be in error,
                please contact one of the human server
                moderators.").queue();
    }
```

这里将针对 Discord 服务器上发布的每条信息，通过“审核端点”和“聊天端点”进行检查。如果任何一个端点都通知我们“该消息已被标记”，我们就会删除频道中的消息，并通知用户他们的消息违反了规则。

现在，内容审核 Discord 机器人变得智能了，让我们试一试！

7.5　运行智能内容审核机器人：ContentModeratorBot.java

现在让我们运行新的改进版内容审核 Java Discord 机器人 ContentModeratorBot.java。执行应用程序后，请务必返回 Discord 服务器并开始提问。图 7-1 展示了机器人的运行情况。

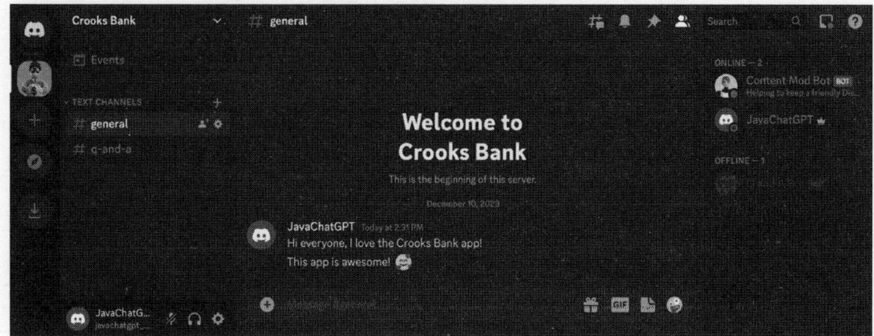

图 7-1 与智能 Discord 内容审核机器人进行讨论：ContentModeratorBot.java

代码清单 7-4 为我们与 Discord 机器人之间的对话，以测试它的能力。

代码清单 7-4 我们与智能审核 Discord 机器人的攻击性对话

Me: Hi everyone, I love the Crooks Bank app!

Me: This app is awesome! 🤩

Me: Come to my website! http://www.google.com

Content Mod Bot: @JavaChatGPT This comment was deemed inappropriate for this channel. If you believe this to be in error, please contact one of the human server moderators.

Me: I'm sorry for breaking the rules. I'm a different person now

Me: But I have some sad news for you

Me: I want to ☠ everyone

Content Mod Bot: @JavaChatGPT This comment was deemed inappropriate for this channel. If you believe this to be in error, please contact one of the human server moderators.

中文注释

我：大家好，我喜欢 Crooks Bank 应用程序！

我：这个应用程序太棒了！🤩

我：来我的网站吧！http://www.google.com

内容审核机器人：@JavaChatGPT 此评论被认为不适合此频道。如果你认为有误，请联系人工服务器审核者。

我：很抱歉违反了规则。我现在是另外一个人了。

我：但我要告诉你一个不幸的消息。

我：我想 ☠ 大家。

内容审核机器人@JavaChatGPT：此评论被认为不适合此频道。如果你认为这是错误的，请联系人工服务器审核者。

在这两个案例中，当有人在 Discord 服务器的任何频道中发布了不当内容时，不仅违规用户会被指出，而且不良信息也会被迅速删除。机器人，真是个得力助手！

你是否注意到，无论是审核端点还是聊天端点，都能读取表情符号？

7.6　小结

在本章中，我们为整个 Discord 服务器创建了一个功能齐全的内容审核机器人！我们利用 OpenAI 的“审核”和“聊天” 端点创建了一个自定义的内容管理器，它不仅可以标记不安全的内容(如仇恨言论和威胁信息)，还可以防止 Discord 服务器的用户受到不必要的骚扰。

7.7　练习

虽然我们在本章中完成了很多工作，但我们还可以做一件事来改进代码。例如进行以下改进：

我们创建的单个 Discord 机器人已经能够识别并避免回复自己发送的消息，但这些机器人还没有学会不重复其他机器人发送的信息。换句话说，如果你同时运行这两个机器人，当有人在“问答”频道中发布了不当内容时，内容审核机器人会立即删除该消息，并通知所有人该消息已被删除。然而，由于技术支持机器人尚不具备识别其他机器人消息的能力，它可能尝试对已经删除的消息进行回复。显然，机器人之间不应该进行这样的对话。因此，下一步的改进方向就是让机器人学会识别其他机器人，并避免做重复工作。

附录A

OpenAI 模型列表

执行代码清单 2-3 中的代码 ListModels.java 后，你将看到一个 JSON 对象，其中包含可供你使用的 OpenAI 模型列表。下表为响应模板。

ID	对象	创建时间	所属权
ada	model	1649357491	openai
ada-code-search-code	model	1651172505	openai-dev
ada-code-search-text	model	1651172510	openai-dev
ada-search-document	model	1651172507	openai-dev
ada-search-query	model	1651172505	openai-dev
ada-similarity	model	1651172507	openai-dev
babbage	model	1649358449	openai
babbage-002	model	1692634615	system
babbage-code-search-code	model	1651172509	openai-dev
babbage-code-search-text	model	1651172509	openai-dev
babbage-search-document	model	1651172510	openai-dev
babbage-search-query	model	1651172509	openai-dev
babbage-similarity	model	1651172505	openai-dev
canary-tts	model	1699492935	System
canary-whisper	model	1699656801	System
code-davinci-edit-001	model	1649880484	openai

(续表)

ID	对象	创建时间	所属权
code-search-ada-code-001	model	1651172507	openai-dev
code-search-ada-text-001	model	1651172507	openai-dev
code-search-babbage-code-001	model	1651172507	openai-dev
code-search-babbage-text-001	model	1651172507	openai-dev
curie	model	1649359874	openai
curie-instruct-beta	model	1649364042	openai
curie-search-document	model	1651172508	openai-dev
curie-search-query	model	1651172509	openai-dev
curie-similarity	model	1651172510	openai-dev
dall-e-2	model	1698798177	system
davinci	model	1649359874	openai
davinci-002	model	1692634301	system
davinci-instruct-beta	model	1649364042	openai
davinci-search-document	model	1651172509	openai-dev
davinci-search-query	model	1651172505	openai-dev
davinci-similarity	model	1651172509	openai-dev
gpt-3.5-turbo	model	1677610602	openai
gpt-3.5-turbo-0301	model	1677649963	openai
gpt-3.5-turbo-0613	model	1686587434	openai
gpt-3.5-turbo-1106	model	1698959748	system
gpt-3.5-turbo-16k	model	1683758102	openai-internal
gpt-3.5-turbo-16k-0613	model	1685474247	openai
gpt-3.5-turbo-instruct	model	1692901427	system
gpt-3.5-turbo-instruct-0914	model	1694122472	system
gpt-4	model	1687882411	openai
gpt-4-0314	model	1687882410	openai
gpt-4-0613	model	1686588896	openai

(续表)

ID	对象	创建时间	所属权
gpt-4-1106-preview	model	1698957206	system
gpt-4-vision-preview	model	1698894917	system
text-ada-001	model	1649364042	openai
text-babbage-001	model	1649364043	openai
text-curie-001	model	1649364043	openai
text-davinci-001	model	1649364042	openai
text-davinci-002	model	1649880484	openai
text-davinci-003	model	1669599635	openai-internal
text-davinci-edit-001	model	1649809179	openai
text-embedding-ada-002	model	1671217299	openai-internal
text-search-ada-doc-001	model	1651172507	openai-dev
text-search-ada-query-001	model	1651172505	openai-dev
text-search-babbage-doc-001	model	1651172509	openai-dev
text-search-babbage-query-001	model	1651172509	openai-dev
text-search-curie-doc-001	model	1651172509	openai-dev
text-search-curie-query-001	model	1651172509	openai-dev
text-search-davinci-doc-001	model	1651172505	openai-dev
text-search-davinci-query-001	model	1651172505	openai-dev
text-similarity-ada-001	model	1651172505	openai-dev
text-similarity-babbage-001	model	1651172505	openai-dev
text-similarity-curie-001	model	1651172507	openai-dev
text-similarity-davinci-001	model	1651172505	openai-dev
tts-1	model	1681940951	openai-internal
tts-1-1106	model	1699053241	system
tts-1-hd	model	1699046015	system
tts-1-hd-1106	model	1699053533	system
whisper-1	model	1677532384	openai-internal